Automated Vehicles: The Role of ISO 26262

Automated Vehicles: The Role of ISO 26262

JUAN R. PIMENTEL

Professor of Computer Engineering
Kettering University

Warrendale, Pennsylvania, USA

400 Commonwealth Drive
Warrendale, PA 15096-0001 USA
E-mail: CustomerService@sae.org
Phone: 877-606-7323 (inside USA and Canada)
Fax: 776-4970 (outside USA)

Library of Congress Catalog Number 2019931392
SAE Order Number PT-206
http://dx.doi.org/10.4271/pt-206

ISBN-Print 978-0-7680-0274-4
ISBN-MediaTech 978-0-7680-0275-1
ISBN-prc 978-0-7680-0278-2
ISBN-epub 978-0-7680-0279-9
ISBN-HTML 978-0-7680-0280-5

To purchase bulk quantities, please contact: SAE Customer Service

E-mail: CustomerService@sae.org
Phone: 877-606-7323 (inside USA and Canada)
Fax: 776-4970 (outside USA)

Visit the SAE International Bookstore at books.sae.org

contents

Introduction

Safety has been ranked as the number one concern for the acceptance and adoption of automated vehicles (AVs) and understandably so since safety has some of the most complex requirements in the development of AVs. The recent fatal accident involving some AVs has made it clear that safety is paramount to the acceptance, testing, verification, and deployment of AVs. ISO 26262 is an important international standard useful for designing automotive systems with a high level of safety. Although ISO 26262 does not address many safety issues of AVs, it is still highly relevant to improve their safety. The systems engineering (SE), risk-based, and lifecycle-based approaches to safety used by ISO 26262 are highly relevant to the design of safe AVs. This book collection will include ten papers covering the implications and use of ISO 26262 in the design of safe AVs.

Index Terms

Automated vehicles

Self-driving vehicles

Autonomous vehicles

Safety

SOTIF

Systems engineering

Control systems

STPA

Functional safety

ISO 26262

I.1 Introduction

Automated vehicles (AVs), also called autonomous or self-driving vehicles, have the potential to reduce accidents, help with the environment, reduce congestion, help the elderly and other disadvantaged populations, and produce other societal benefits [1, 2, 3, 4, 5]. However, the touted advantages of autonomous vehicles and those including latest advanced driver assistance system (ADAS) features are turning out difficult to sell to the public than many manufacturers and tier 1s have anticipated. Much of the early euphoria of self-driving vehicles is diminishing in the wake of some recent accidents involving AVs with varying degrees of automation. A recent online marketplace for buying and selling cars found 69% of respondents are scared of autonomous automobiles. It is also found that these people found technology in cars helpful (58%), but only 12% said ADAS and infotainment features were a "must-have." The survey asked more than 1,000 respondents from across the United States geographically and across age groups, although it should be noted the biggest group was 60+ years old*. Accidents involving self-driving vehicles are inevitable; as it is the case with other industries, accidents have happened and will happen no matter the efforts made to avoid them. The National Transportation Safety Board (NTSB) has issued a report† on a Tesla accident on May 7, 2016, and two preliminary reports on an Uber accident‡ on March 18, 2018, and a Tesla accident§ on March 23, 2018. After analyzing these reports, what is disturbing are the details associated with these accidents which indicate that as an industry, we may need to go back to safety 101.¶

Regarding the aforementioned accidents, it would not be so bad if the safety systems of the vehicles in question were designed and functioning properly according to their stated automation level. In the case of the Tesla accident in Florida, the vehicle failed to activate the forward collision warning (FCW) system and automatic emergency braking (AEB). In the case of the Uber accident, emergency braking maneuvers were not enabled while the vehicle was under computer control, to reduce the potential for erratic vehicle behavior, and the system relied on the vehicle operator for safety. In the Tesla accident in California, the vehicle failed to detect a damaged crash attenuator and hit it at a speed of about 71 mph.

After recent incidents and mishaps involving AVs such as those described above, it is clear that there is much room for improvement not only by manufacturers but also by government regulations,

* http://analysis.tu-auto.com/autonomous-car/shifting-public-acceptance-autonomous-tech?NL=TU-001&Issue=TU-001_20180723_TU-001_235&sfvc4enews=42&cl=article_2_2.

† https://www.ntsb.gov/investigations/AccidentReports/Reports/HAR1702.pdf.

‡ https://www.ntsb.gov/investigations/AccidentReports/Reports/HWY18MH010-prelim.pdf.

§ https://ntsb.gov/investigations/AccidentReports/Reports/HWY18FH011-preliminary.pdf.

¶ https://www.eetimes.com/document.asp?doc_id=1333446&_mc=RSS_EET_EDT&utm_source=newsletter&utm_campaign=link&utm_medium=EETimesWeekInReview-20180721.

researchers, the general public, and other stakeholders. Over the past few months, the media has been full of headlines such as "How Safe Is Driverless Car Technology, Really?", "Autonomous Cars: How Safe Is Safe Enough?", and "How safe should we expect self-driving cars to be?" In addition, some industry analysts and safety experts are offering advice to tech and automotive companies to reconsider their safety programs. There is also some agreement that "the self-driving car industry's reputation has suffered a setback," and the question is how to fix it.* It appears that AV companies are much more stringent when using semiconductor devices and EDA tools demanding that they conform to ISO 26262 than using the same yardstick for their own safety-critical designs.

So what is there to do? Safety is not new; at least for the last 60 years, it has been successfully applied in several industries such as nuclear, avionics, process control, automotive, and others. What is unique and special about the safety of self-driving vehicles? What should be the emphasis for a more effective AV safety program? What are the roles of governments, standards, testing, verification, validation, and sound safety engineering efforts? Addressing issues regarding autonomous vehicle safety is challenging [7]. Currently, as an industry, we just do not fully understand the nature of self-driving vehicle safety and how to design safe AVs. For example, there is little discussion on ways to estimate, analyze, compute, or measure the level of safety of an AV design or AVs. We need to begin by fully characterizing it and this book series is an effort in this direction.

Some manufacturers such as Waymo cite their recent milestone of 8 million miles driven on public roads as a measure of the safety achieved by their self-driving vehicles.† However, it is not clear how a certain number of millions of miles driven contribute to the safety level of self-driving vehicles. Some industry analysts believe that policy makers and city officials overseeing infrastructure will be the most important players in reshaping the self-driving vehicle safety landscape. For example, the National Highway Traffic Safety Administration (NHTSA) has issued a voluntary guidance whose purpose is to help designers of automated driving systems (ADSs) analyze, identify, and resolve safety considerations prior to deployment using their own, industry, and other best practices.‡ It outlines 12 safety elements, which the agency believes represent the consensus across the industry, that are generally considered to be the most salient design aspects to consider and address when developing, testing, and deploying ADSs on public roadways. Within each safety design element, entities are encouraged to consider and document their use of industry standards, best practices, company policies, or other methods they have employed to provide for increased

* https://www.eetimes.com/document.asp?doc_id=1333446&_mc=RSS_EET_EDT&utm_source=newsletter&utm_campaign=link&utm_medium=EETimesWeekInReview-20180721.

† https://www.theverge.com/2018/7/20/17595968/waymo-self-driving-cars-8-million-miles-testing.

‡ https://www.nhtsa.gov/sites/nhtsa.dot.gov/files/documents/13069a-ads2.0_090617_v9a_tag.pdf.

system safety in real-world conditions. The 12 safety design elements apply to both ADS original equipment and replacement equipment or updates (including software updates/upgrades) to ADSs. However the NHTSA guidance is not specific enough to help manufacturers in designing effective safety mechanisms to reduce risk.

I.2 Characterizing the Safety of Automated Vehicles

How different is the concept or notion of safety in self-driving vehicles when compared to that used in other industries such as aviation, process control, and automotive? While the fundamental concepts are the same, the safety of self-driving vehicles has specific attributes that are different or not present in the safety of other industries. In this section we briefly discuss these attributes. When compared to the safety of traditional industries such as avionics, process control, and automotive, there are specific attributes pertaining to the safety of self-driving vehicles that we discuss next [23].

I.2.1 Performance Degradation

Traditional safety is based on faults and failures of mostly hardware components, and this is referred to as the reliability approach to safety [9]. In contrast, accidents involving self-driving vehicles might happen even if no hardware device fails, but rather a performance degradation of some of its functions or intended functionality occurs. Addressing safety issues for these situations is referred to as safety of the intended functionality (SOTIF), and it is a fairly new concept as applied to the safety of self-driving vehicles [10]. Thus, some failures are due to performance degradation of self-driving vehicle components, typically involving higher levels of processing or higher levels of automation, for example, service failures. This definition of failure goes beyond that which is defined in the standard ISO 26262; however, it is compatible with other safety frameworks such as [8], System Theoretic Process Analysis (STPA) [11, 12, 13, 14, 15], or other real-time distributed systems [16]. One example of this understanding of the concept of safety is the failure of a vehicle detection system where the perception system provides missed detections (i.e., false negatives) or spurious detections (i.e., false positives). This could happen because the processing of environmental data is highly complex and the object detection function is subject to errors and impairments, particularly in bad weather or in night conditions when the visibility is poor. Either one of these failures could be catastrophic and could result in an accident or harm. Another example is a radio detection and ranging (RADAR) system correctly detecting objects only when the objects are moving, thus missing static objects because of limits on its performance. Thus, failure occurs in a degraded performance scenario.

I.2.2 Focus on Software

It is well known that the amount of software in a vehicle continues a rapidly increasing trend that started with the development of by-wire systems. Much of the functionality of a self-driving vehicle is implemented in software, and thus it is important to view the perception system as a set of software servers each providing services to the rest of the system. Therefore, one can refer to these various functionalities as a vehicle detection server, a pedestrian detection server, a road detection server, etc. The software in self-driving vehicles is much larger in size and scope compared to traditional industries; thus, there should be a focus on the safety of the software. As noted, a failure can occur if the software services deviate from the correct services, and this could lead to safety hazards and safety risks. Ultimately, the overall safety of a self-driving vehicle will be dictated by the safety of its software [9, 12].

I.2.3 Non-Deterministic Perception System

In the absence of hardware faults, the perception systems of traditional industries are mostly deterministic in nature. For example, sensing the intake manifold pressure or engine speed in automotive systems is deterministic.[4] In contrast, the perception systems of self-driving vehicles are non-deterministic, leading to a high level of false positives and false negatives when their performance deteriorates to the point that service failures cannot be avoided. The non-deterministic aspect of the perception system stems from the fact that one never knows when its performance will deteriorate to the point where failures begin to appear in the services delivered by the system. Thus, the services provided by the perception system are subject to random failures, for example, when the weather deteriorates or when the system makes detection errors.

I.2.4 Perception System Complexity

Sensing elementary physical phenomena such as temperature or pressure is relatively simple, involving just some deterministic sensors, some electronics, and communications. In contrast, sensing or detecting man-made entities or constructs such as another vehicle, a road boundary, a city street, or a street intersection is complex because of the lack of structure of what is being sensed or perceived. The implication of the complexity of the perception system is that it is prone to errors, which degrades the performance or safety of the overall vehicle.

I.2.5 Overall System Complexity

In addition to its perception system, a self-driving vehicle also includes localization and mapping, planning and control, and actuation resulting in a highly complex system. One of the main issues with system complexity is that it makes testing for safety challenging, particularly if machine learning techniques are used, as it makes the design opaque to humans.

This makes tracing the design and the test plans to the requirements problematic, since there is no human-understandable design that can be used for verification and testing [17]. In addition, it is known that when the system is complex, the system safety is affected by interacting complexity and tight coupling [9]. Another aspect of system complexity is that the autonomous vehicle operates in a complex external environment, and there are safety hazards due to events outside the domain of the autonomous vehicle, for example, from other vehicles (whether self-driving or not). Thus, the safety attributes of a self-driving vehicle are significantly different from those in other industries such as avionics, process control, and automotive.

In addition to its attributes, what are the various types of safety that encompass the overall safety of self-driving vehicles? As noted, the safety of self-driving vehicles is complex and differs from that of other industries such as avionics, process control, and automotive. On the one hand, there are safety commonalities such as the safety that involves component failures, which is the subject of so-called functional safety, and the safety involving components whose failure rates are well understood because they are proven in use, that is, in actual operation. On the other hand, there are two types of safety that are not prevalent in the avionics, process control, and automotive industries, and these include *SOTIF* and *multi-agent safety*. Thus, the types of safety that characterize the safety of self-driving vehicles include (1) traditional functional safety as defined by ISO 26262, (2) SOTIF, and (3) multi-agent safety [23]. Feth et al. also emphasize that safety assurance is a concern because established safety engineering standards and methodologies are currently not sufficient [20, 21, 22]. They also conclude that there are three types of safety that characterize the safety of self-driving vehicles: (1) traditional functional safety, (2) SOTIF which they assume are due to *functional insufficiencies*, and (3) multi-agent safety related to safe driving behaviors which are abstracted from technological challenges of situation awareness. Furthermore, they elaborate the fundamental safety engineering steps that are necessary to create safe vehicle of higher automation levels while mapping these steps to the guidance presently available in existing (e.g., ISO 26262) and upcoming (e.g., ISO PAS 21448 [26]) standards. Functional safety is a well-understood area which is guided by a number of international standards such as IEC 61508 [18], IEC 61511 [19], and ISO 26262 [6], and there are a large number of papers and publications on this topic. However, it is noted that ISO 26262 does not cover AVs; thus, its application should be done with great care [27, 28, 29, 30, 31]. In the following, we characterize the safety category of SOTIF.

I.3 Role of Systems Engineering

Many systems and products are very complex, very large, or both; certainly one example is an AV with a high degree of safety, thus the complexity of designing and developing an autonomous vehicle with a high degree of safety. Systems engineering (SE) is a methodology that includes a set of processes, techniques, and procedures to design systems that are very complex and/or very large. SE is based on the concept of a system lifecycle composed of a number of phases such as concept, product development, and production and operation. The concept phase includes many sub-tasks such as item definition, initiation of safety lifecycle, hazard analysis and risk assessment (HARA), and functional safety concept (FSC). The product development phase can be performed at a system, hardware, and software levels. Finally, the production and operation (or utilization) phase includes production, operation, service (maintenance, repair, etc.), and decommissioning.

The number and names of system lifecycle phases are not unique. Other phases identified with a system lifecycle include requirements phase, preliminary and detailed design phase, construction and production phase, and the operation phase. The requirements phase is an important phase that takes place at the beginning of a project that involves top-level requirements at all functional levels. Some requirements are further developed or extended into "derived requirements." This process is repeated appropriately until reaching the lowest level possible. There are two types of requirements, functional and non-functional. Requirements need to be carefully selected in order to ensure that they make sense in the context of the final outcome of the project and conveyed to all the team members. Missing out on a requirement or misapplying one could spell disaster for a project. A functional requirement specifies what the system should do, that is, it describes a particular behavior or function of the system when certain conditions are met, for example: "Send email when a new customer signs up" or "Open a new account." A functional requirement for an everyday object like a cup would be: "ability to contain tea or coffee without leaking." Examples of some functional requirements include: carry people on public streets and roadways, carry people through air space, provide external interfaces, define business rules, perform transaction corrections, perform adjustments and cancellations, provide administrative functions, provide authentication, use authorization levels, perform audit tracking, etc. A non-functional requirement specifies how the system performs a certain function, that is, it describes how a system should behave and what limits there are on its functionality; it specifies the system's quality attributes or characteristics, for example: "Modified data in a database should be updated for all users accessing it within 2 seconds." A non-functional requirement for the cup mentioned previously would be: "contain hot liquid without heating up to more than 45 °C." Important categories of non-functional requirements include availability, safety, security, and reliability. The preliminary and detailed design phase includes intermediate and final designs with enough details and specifics for its implementation by other parties. It is based on requirements and involves assumptions, calculations, measurements, simulations, etc. The outcomes of this phase are produced using an assortment of tools in the categories of drawing, simulation, SE, etc. The construction and production phase is that where the system is built up using blueprints of the detailed design phase. When the construction is done in series, for example, in a

FIGURE 1 Systems engineering V model listing main tasks on the left and right sides of the V.

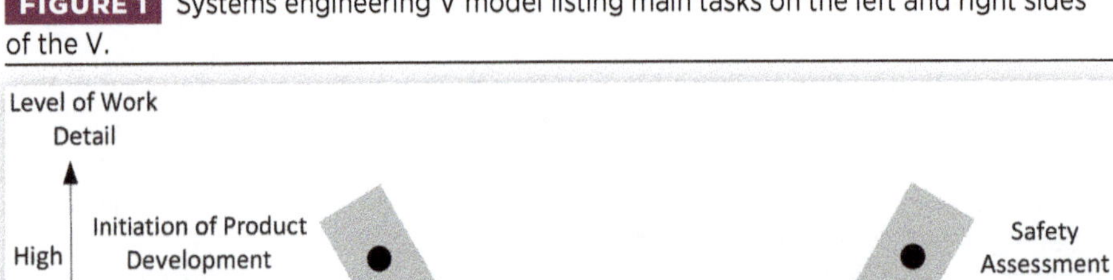
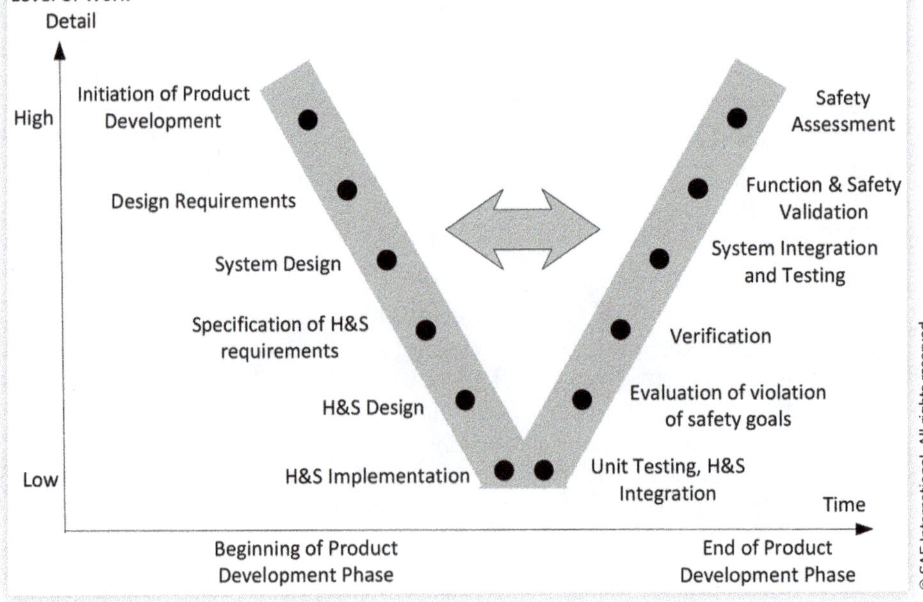

manufacturing plant, then it is termed production. Testing is an important activity to be done during construction. There are many types of tests such as unit test, system test, interoperability test, stress test, system test, etc. The operation phase starts when the system or product is put in service and ends when the system is taken out of service or decommissioned. It includes the following activities: operation, maintenance, troubleshooting, repair, etc.

SE also uses the so-called V model to specify and describe key tasks while designing a complex or large system. These key tasks include requirements, design, implementation, test, verification, and validation which are performed at specific times and in a *V* shape depicted in Figure 1. The lifetime phases (or tasks) of requirements, design, and implementation are performed on the left side of the V, while the tasks involving testing, verification, and validation are performed on the right side of the V. The reason for depicting these activities on a V (rather than in a linear fashion) is that while working on the requirements specifications, the corresponding specifications for validation are developed concurrently and correspond to the same vertical level. While the completion of the requirements specification task is made very early in the development lifecycle of the product or project, the completion of the validation task is the last one. On the left side of the V, decomposition and definition activities resolve the system architecture and create the details of the design. Integration and verification flow up and to the right as successively higher levels of subsystems are verified, culminating at the system level. Verification ensures the system was built right (meets requirements, standards, etc.), whereas validation ensures the right system was built (meets customers' needs). Ascending the right side of the V is the process of integration and verification, and at each level, there is a direct correspondence between activities on the left and right sides of the V. This correspondence is deliberate—the method of verification must be determined as the requirements are developed and documented at each level. This minimizes the chances that requirements are specified in a way that cannot be measured or verified.

Although SE can be applied to any endeavor (e.g., physical and social sciences, engineering, and management, etc.), in this section we discuss its application to engineering and, more specifically, designing a safe AV. One of the most fundamental ideas of SE is the following. At each level, starting at highest level of abstraction and proceeding to lowest, identify main functions using input/output relationships and use functional decomposition techniques to generate sub-functions at lower levels of abstraction. This decomposition process is repeated until four to eight levels (typically) deep are generated. At each level sufficient details are added to clarify the composition, relationships, requirements, behaviors, interfaces, parameters, etc. This process is illustrated in Figure 2 (a) and (b) where a system or sub-subsystem is decomposed into constituent components until a subsystem with readily available components can be implemented.

Designing a vehicle with a high level of safety is complex, thus the importance of using SE principles. Indeed, the ISO 26262 standard which has been specifically developed for automotive safety uses the V model as the underlying framework. This decision minimizes unnecessary rework and errors in requirements development and cascading. This will force some testing and verification activities to occur at various levels of integration early in the process. More specifically, the ISO 26262 standard uses the V model at the system level (Section 4), at the hardware level (Section 5), and at the software level (Section 6).

FIGURE 2 (a) Top-level block diagram of a system.

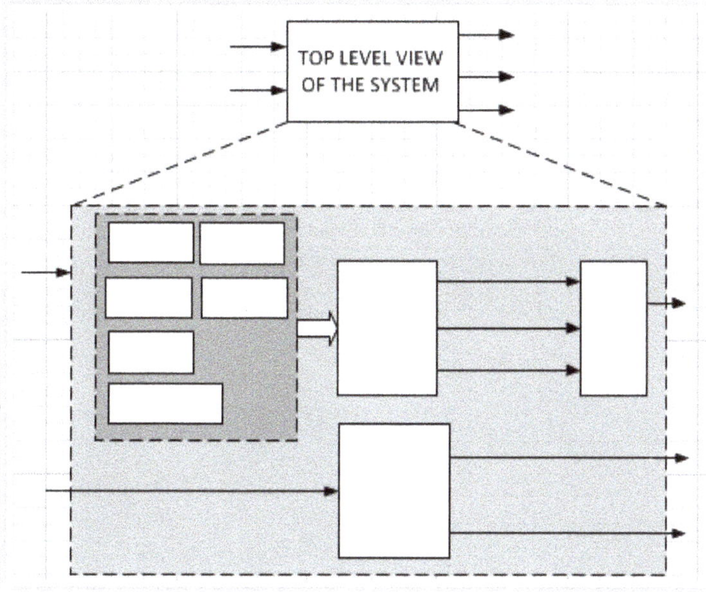

TOP LEVEL VIEW OF THE SYSTEM

FIGURE 2 (b) Functional decomposition of a component into its constituent sub-components.

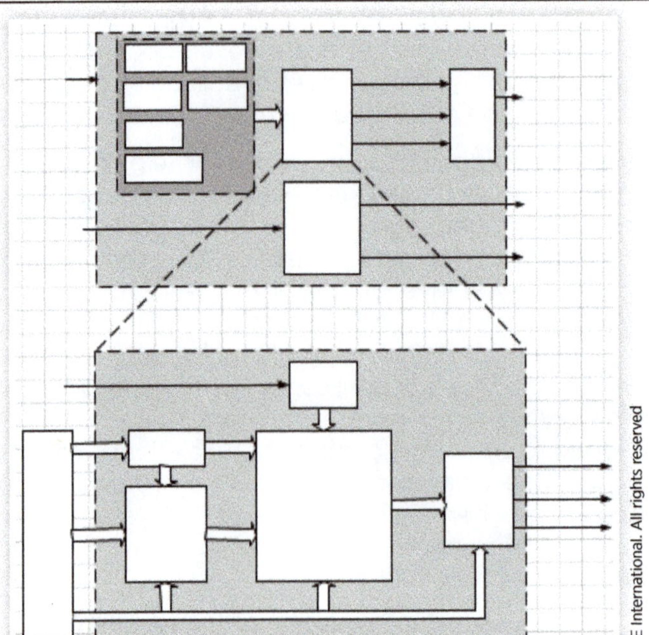

1.3.1 Model-Based Systems Engineering (MBSE)

Model-based systems engineering (MBSE) is a SE methodology that focuses on creating and exploiting domain models as the primary means of information exchange between engineers, rather than on document-based information exchange. These models fall into many categories including functional, behavioral, structural, operational, component, performance, safety, and many others. The benefits that are attributed to using an MBSE approach include shared understanding of system requirements and design, assisting in managing complex system development, improving design quality, supporting early and ongoing verification and validation to reduce risk, providing value through lifecycle, and enhancing knowledge capture. One effective way to benefit from MBSE is through specific modeling languages for SE which have incorporated many domain models in an intrinsic fashion as is the case with Systems Modeling Language (SysML) which is summarized next.

1.3.2 The Systems Modeling Language

SysML is a general-purpose modeling language for SE applications. It supports the specification, analysis, design, verification, and validation of a broad range of systems and systems of systems. These systems may include hardware, software, information, processes, personnel, and facilities. SysML is becoming important in the execution of many safety tasks associated with the V model of SE. There are four major categories of SysML diagrams, also known as the "four pillars of SysML": structure, behavior, requirements, and parametrics. The structure category basically allows one to describe what the system is including its main constituent components. While this is important to get started, the structure category is static as it does not describe how the system works or behaves which is the function of the behavior category. Thus the behavior category describes the dynamic aspect of the system. As noted, all design works begin with a set of requirements describing what one can expect from the system. The requirements category of SysML enables the specification of these requirements. A system typically consists of many subsystems with a number of interfaces. The parametrics category defines the nature of these interfaces including quantification beyond qualitativeness, for example, formulas, equations, units, etc.

In most cases, just nine SysML diagrams are used which belong to the four categories described previously. The structure category includes three diagrams: block definition diagram (*bdd*), internal block diagram (*ibd*), and package diagram (*pkg*). Four diagrams belong to the behavior category: sequence diagram (sq), state machine diagram (*stm*), activity diagram (*act*), and use case diagram (*uc*). The requirements and parametrics categories each include just one diagram, the requirement diagram (*req*) and parametric diagram (par), respectively. In the following, we briefly summarize some of these SysML diagrams.

The SysML bdd represents system elements called blocks and their composition, classification, and navigation. The following figure illustrates the *bdd* of a perception system of an AV (Figure 3).

The SysML ibd represents interconnection and interfaces between the parts of a block including external ports of the block. The following figure illustrates the ibd of a perception system of an AV (Figure 4).

The SysML pkg represents the organization of a model in terms of packages that contain model elements. Thus this diagram is used for model structuring rather than system structuring. The SysML sd represents behavior in terms of a sequence of messages exchanged between parts. The following figure illustrates the sd of a RADAR sensor subsystem pertaining to the perception system of an AV (Figure 5).

The SysML stm represents behavior of an entity in terms of its transitions between states triggered by events. The following figure illustrates the stm of an antilock braking system (ABS) of an automobile (Figure 6).

The SysML act represents behavior in terms of the ordering of actions based on the availability of inputs, outputs, and control and how the actions transform the inputs to outputs. The following figure illustrates the act of an ABS of an automobile (Figure 7).

The SysML uc represents functionality in terms of how a system or other entity is used by external entities (i.e., actors) to accomplish a set of goals.

FIGURE 3 The block definition diagram (bdd) of SysML.

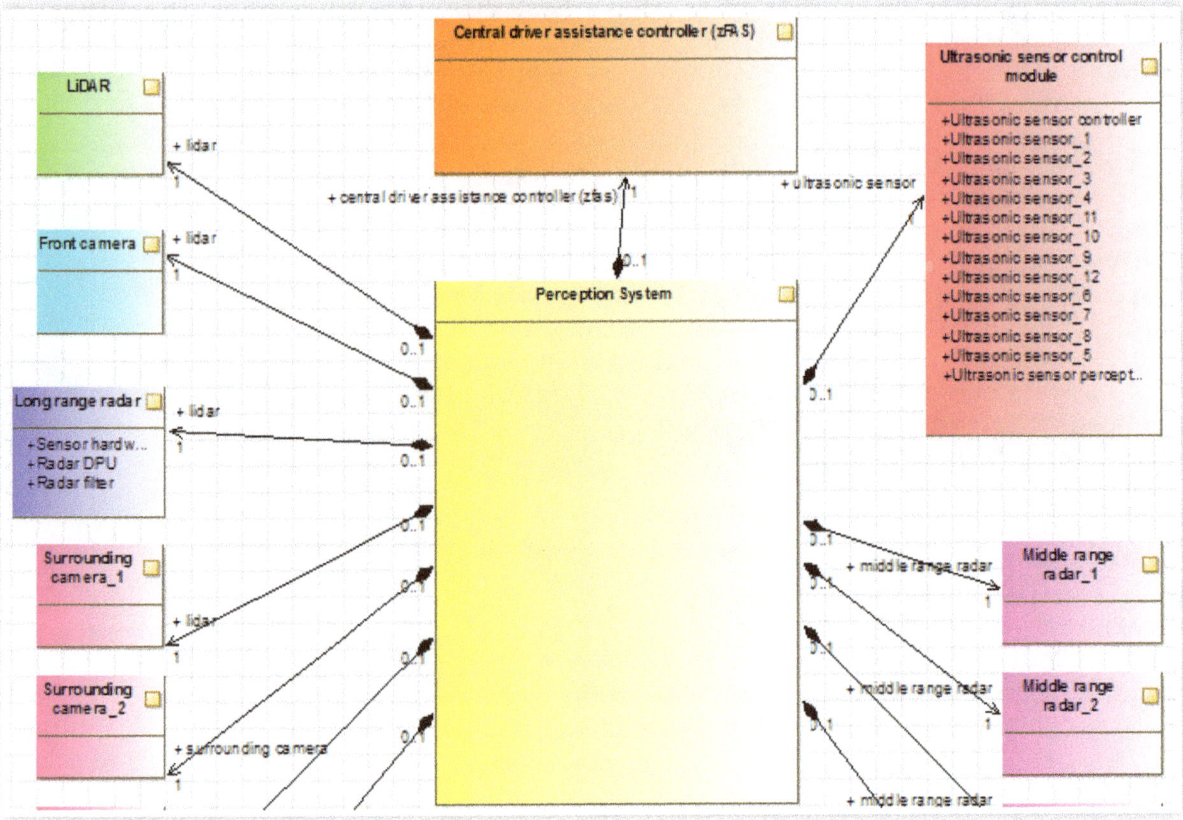

Central driver assistance controller (zFAS)

LiDAR

+ lidar

1

+ central driver assistance controller (zfas)

1

+ ultrasonic sensor

Ultrasonic sensor control module

+Ultrasonic sensor controller
+Ultrasonic sensor_1
+Ultrasonic sensor_2
+Ultrasonic sensor_3
+Ultrasonic sensor_4
+Ultrasonic sensor_11
+Ultrasonic sensor_10
+Ultrasonic sensor_9
+Ultrasonic sensor_12
+Ultrasonic sensor_6
+Ultrasonic sensor_7
+Ultrasonic sensor_8
+Ultrasonic sensor_5
+Ultrasonic sensor percept...

Front camera

+ lidar

0..1

Perception System

Long range radar

+Sensor hardw...
+Radar DPU
+Radar filter

+ lidar

Surrounding camera_1

+ lidar

Surrounding camera_2

+ surrounding camera

Middle range radar_1

+ middle range radar

Middle range radar_2

+ middle range radar

+ middle range radar

FIGURE 4 The internal block diagram (ibd) of SysML.

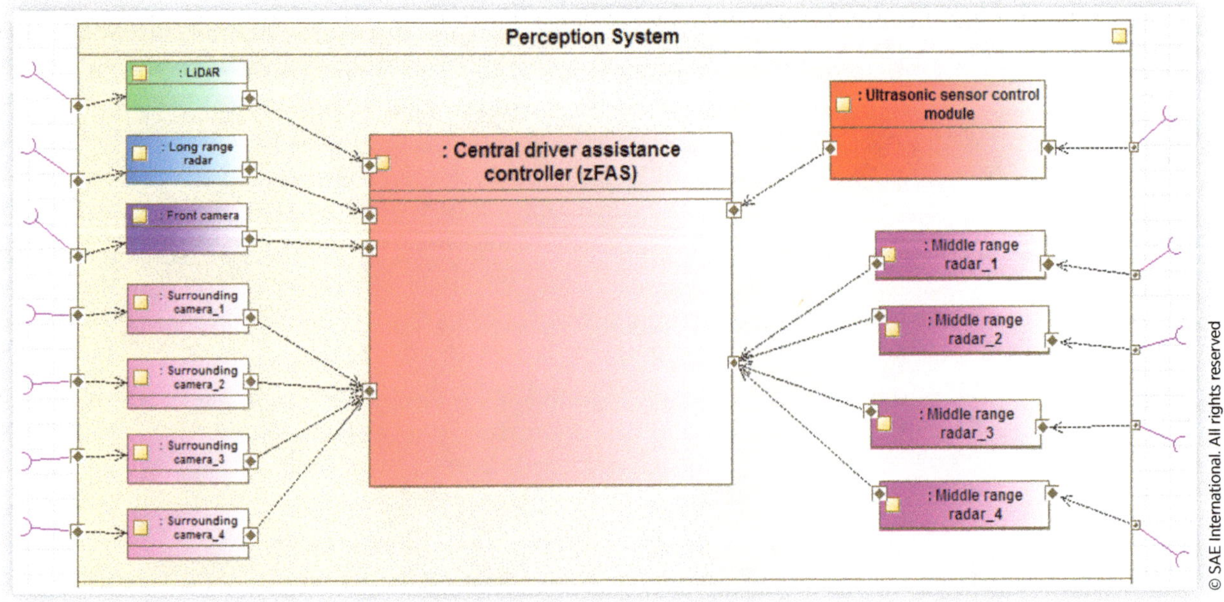

FIGURE 5 The sequence diagram (sd) of SysML.

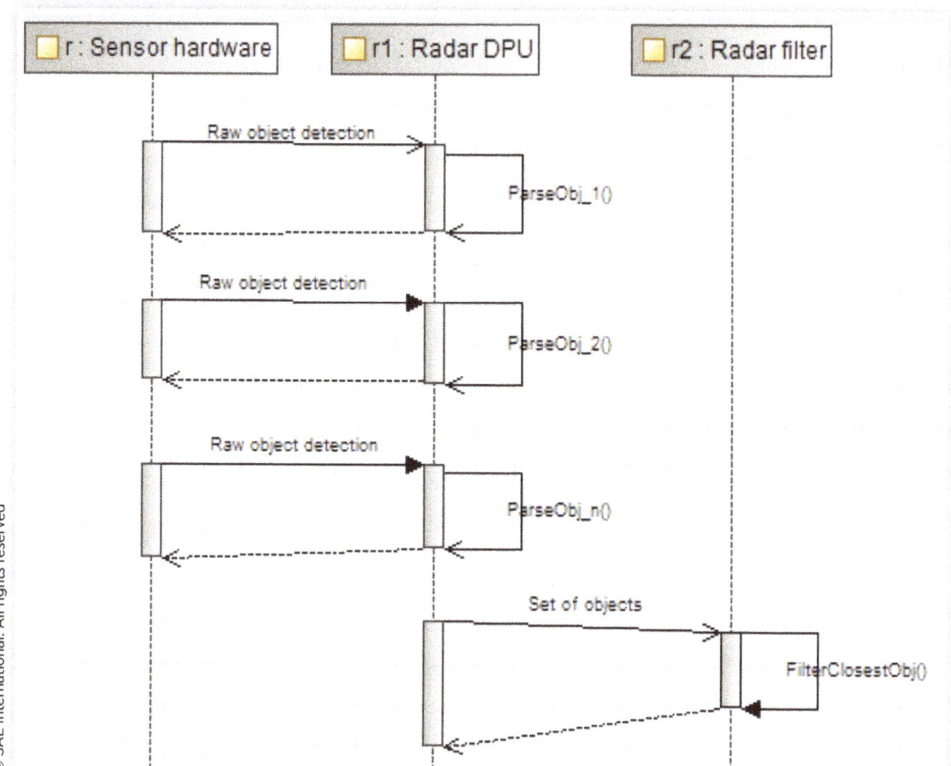

FIGURE 6 The state transition diagram (std) of SysML.

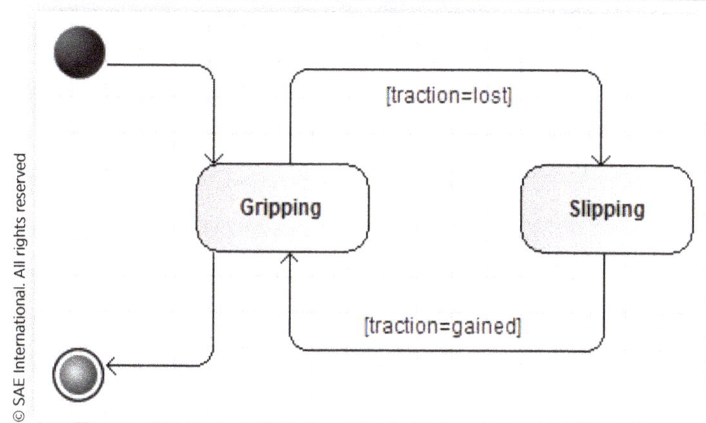

FIGURE 7 The activity diagram (act) of SysML.

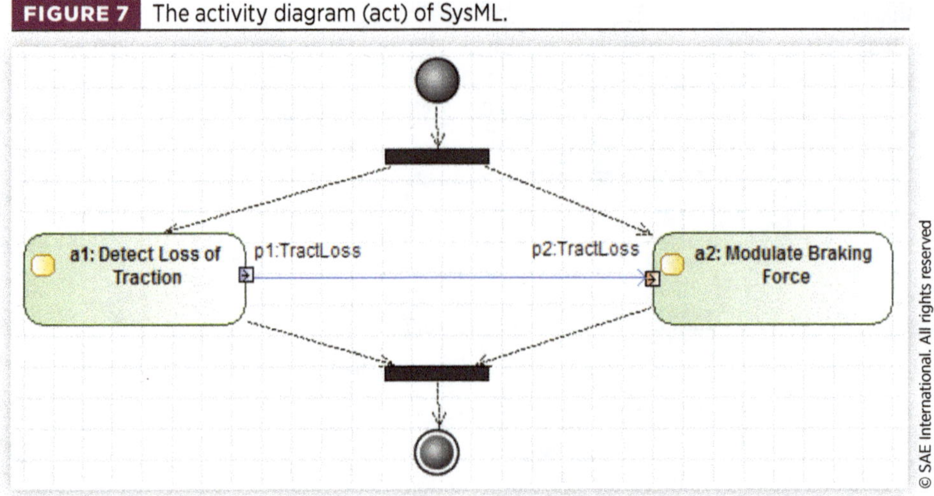

FIGURE 8 The requirements diagram (req) of SysML.

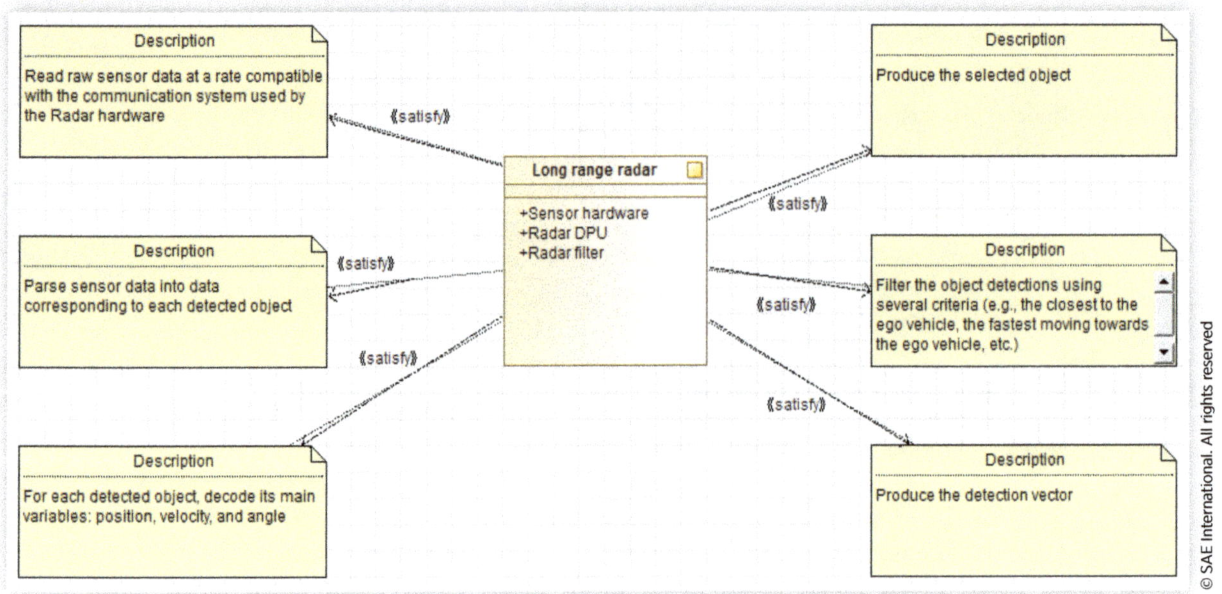

The SysML req represents text-based requirements and their relationships with other requirements, design elements, and test cases to support requirements traceability. The following figure illustrates the req of a RADAR sensor subsystem pertaining to the perception system of an AV (Figure 8).

The SysML *par* represents constraints on property values, such as $F = m*a$, used to support engineering analysis. The following figure illustrates the *par* of an ABS of an automobile (Figure 9).

FIGURE 9 The parametric diagram (par) of SysML.

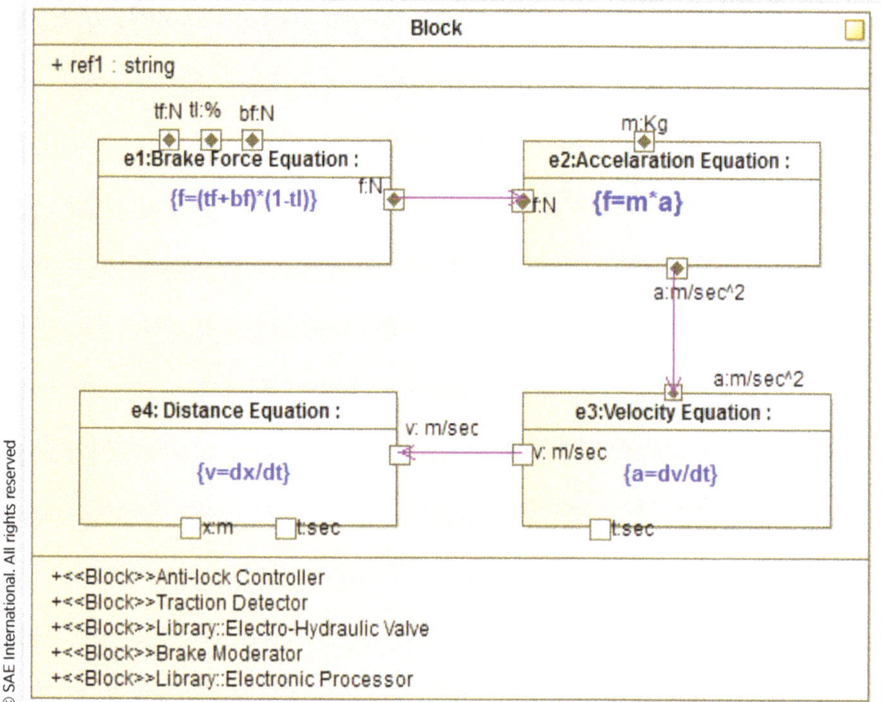

I.4 Control Systems

Control systems are pervasive in automotive and AVs, thus its importance to view them with a proper perspective. A control system maintains an output according to a prescribed fashion called reference signal or set point regardless of any disturbances or other dynamic conditions. The main components of a control system include the process (or plant) to be controlled, the controller, sensors, and electronic or computing elements for its implementation. For example, the motion control system of an AV maintains the trajectory of the vehicle according to a pre-calculated path regardless of disturbances. Likewise, the cruise control system of a vehicle maintains the speed of the vehicle in a pre-specified fashion regardless of disturbances.

There are many types or categories of control systems and associated controllers and a short list include:

- Open-loop control
- Closed-loop or feedback control loop
- Cascading control
- Model predictive control (MPC)
- Optimal control

FIGURE 10 Basic feedback control system also called a feedback control loop.

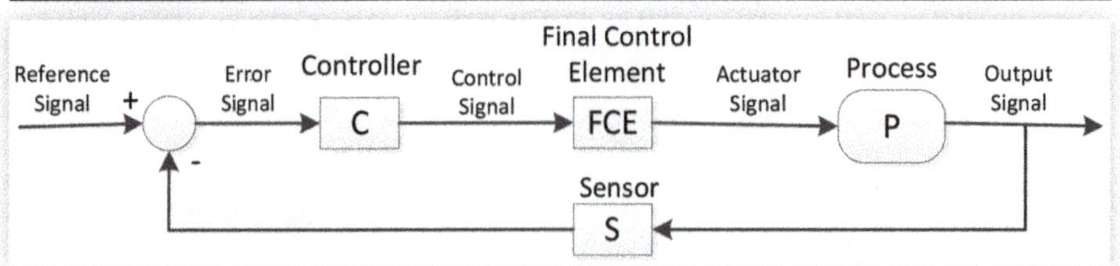

The open-loop control system is the only one in the category that does not use feedback for control purposes, that is, it does not use sensors to determine the values of the outputs of the control system. Thus, its application is very limited and we do not discuss them any further. The closed-loop or feedback control system is by far the most used, and it can exist by itself or as a component of other more sophisticated control system (see Figure 10). The most important variables in a feedback control system include the reference signal (or set point), the error signal, the control signal, the actuator signal, and the output signal, also called the controlled variable. Feedback involves data or information about the status (or the magnitude) of the controlled variable which can be compared with its reference, desired value, prescribed value, or set point. In a feedback control system, the outcome of an action (i.e., the output signal) is fed back and compared with the reference signal, and the error signal is used by the controller for corrective action through control signals. These control signals do not usually have the power to effect change on the process; thus an actuator or final control element (FCE) is necessary as shown in the figure. The main advantages of the feedback control signal are that it is a relatively simple design when compared with other types of control systems and it is relatively easy to tune (for linear and near-linear systems). Its disadvantages include slow recovery for slow processes upon load disturbance or change in set point. A process with long dead time may exhibit oscillatory process response and with longer settling times (Figure 10).

A cascade control system consists of several control systems organized in a hierarchical fashion. Although there can be a number of levels, we will illustrate one that consists of just two levels, an outer level control loop and an inner level control loop as illustrated in Figure 11 where the inner control loop is made up of elements C2, FCE, P2, and S2. Likewise, the outer control loop is made up of elements C1, P1, and S1 and the entire inner control loop. In the cascade (or cascading) control system,

FIGURE 11 Cascade control system consisting of two feedback control loops, an inner and an outer.

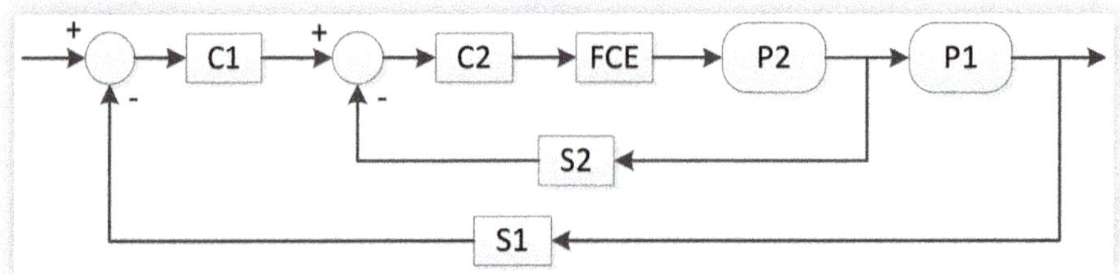

FIGURE 12 Hierarchical control systems corresponding to an automated vehicle.

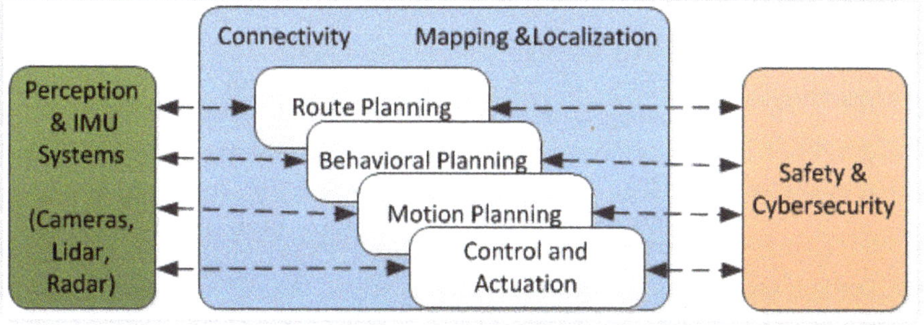

the inner loop controller obtains its set point from the outer loop controller, that is, the reference signal for the inner loop control system is the output of controller C1. The purpose is to implement hierarchical control systems and to reduce or eliminate the effect of disturbances. A good example is the cascading control system in an AV where the various feedback loops correspond to the following control levels as depicted in Figure 12: route planning, behavioral planning, motion planning, and control and actuation. Advantages of cascading controllers include the implementation of hierarchical controllers and faster recovery time as compared to single feedback loop. One disadvantage is that the controllability of the system will be worse than a single feedback loop if both controllers are not properly tuned or calibrated.

As noted, control systems are pervasive in an AV. In the context of Figure 12, we have the following control systems involving route planning controller, behavioral planning controller, motion planning controller, and lowest-level controller and actuator. All these controllers operate in a cascading fashion, that is, the output of a higher-level control constitutes the reference input for the next lower-level controller. More specifically, the route planning controller outputs a route that becomes the reference input for the behavioral control system which in turn outputs a trajectory that becomes the reference input for the motion planning control system. The trajectory generated by the motion planning controller may need to be modified because of other vehicles or other objects that might be blocking the trajectory. By using the perception system and operating in a real-time fashion, the motion planning control system generates a path to be followed by the vehicle. Finally, the lowest-level control system takes this path as its reference input and generates the actual signals for the vehicle actuators (corresponding to steering, acceleration, and braking) so that the vehicle follows the reference path. In the context of the cascading control system of Figure 11, C2 and FCE constitute the *control and actuation* level in Figure 12, and controller C1 corresponds to the *motion planning* controller in Figure 12. Process P2 corresponds to the vehicle propulsion and steering, process P1 represents the environment external to the vehicle, sensor S1 corresponds to the vehicle perception system, and sensor S2 corresponds to the sensors associated with the vehicle propulsion and steering systems.

1.5 STPA

One of the critical tasks of designing a safe AV is the identification of safety hazards and risks including safety measures and mechanisms to reduce risk, and a number of techniques are available to do this [24, 25]. One such technique is DFMEA (design failure modes and effect analysis), which is primarily based on component faults which can lead to failure modes such as hole too shallow, hole too deep, hole missing, hole off-location, dirty, deformed, surface finish too smooth, burred, open circuited, short circuited, bent, cracked, misaligned, missing label, etc. The Society of Automotive Engineers (SAE) standard J1739 is a reference guide that covers potential failure mode and effects analysis (FMEA) in design, manufacturing, assembly processes, and machinery.

Another technique for the identification of hazards and risks is STPA which is based on SE and control systems. This is in contrast to DFMEA which is primarily based on component failures. Although the J1739 standard does take into account interfaces, interactions, and system architecture in a generic fashion, however, the viewpoint is not from a SE or control system perspective as it the case with STPA. Thus the goals of DFMEA and STPA are similar; however their approaches are widely different. STPA has been applied to address safety issues in AVs from various perspectives that include requirements development, system architecture, software, verification, and others. For example, the casual factors to create scenarios for hazard identification in a basic feedback loop are shown in Figure 13.

1.5.1 HARA and Risk Reduction

Both SAE J1739 and STPA can be used to perform HARA and risk reduction of AVs with the proper interpretation of the standard and methodology, respectively. In the case of STPA, the process begins with identifying hazards in a control system framework and related system-level

FIGURE 13 Causal factors to create scenarios for hazard identification in STPA.

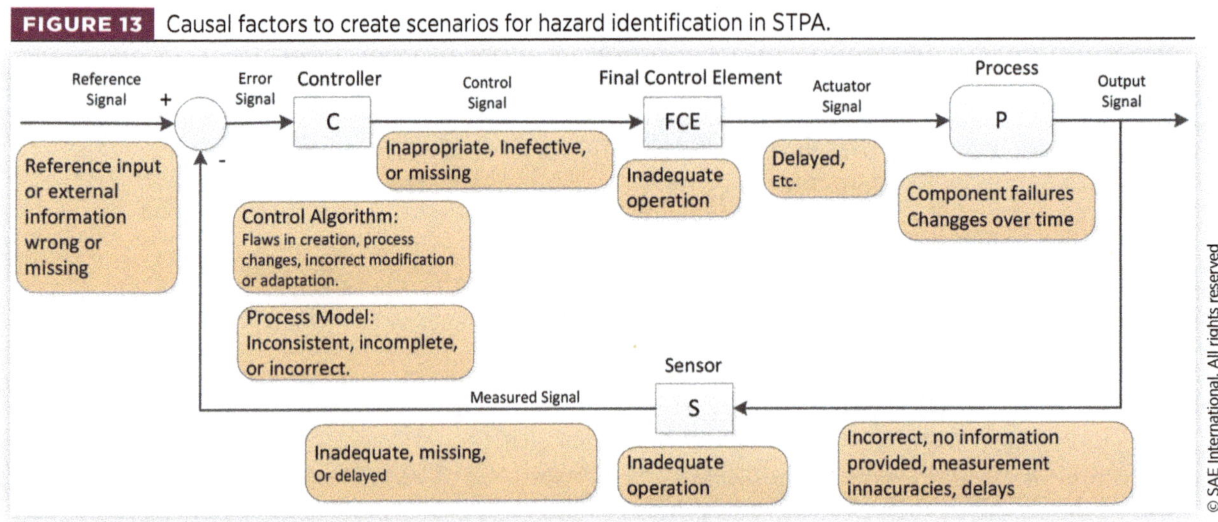

assumptions. The process is followed by two important steps: STPA Step 1 and STPA Step 2. In STPA Step 1, a set of undesired control actions (UCAs) are defined and a set of safety constraints are developed. UCAs are identified and classified into four types: (a) a control action required is not provided or not followed, (b) an UCA is provided, (c) a control action is provided too early or too late, and (d) a control action is stopped too soon or applied too late. After UCAs are identified, STPA Step 2 identifies causal factors and scenarios that potentially lead to UCAs. These are based upon an accident causality model using physical principles. Finally, STPA Step 2 includes mitigation solutions to reduce risk. These mitigation solutions are facilitated by performing the many detailed steps and processes of STPA.

I.6 The Papers in This Collection

There is a fair amount of papers in the areas of SE, control systems, and the role of ISO 26262 in the development of AVs from SAE and other sources. The following is a representative list of ten SAE papers on the subject.

1. SuSanta Sarkar and James Forsmark, "System Engineering for Automated Software Update of Automotive Electronics," SAE Technical Paper 2018-01-0750, 2018, doi:10.4271/2018-01-0750.
2. Andreas Himmler, Klaus Lamberg, Tino Schulze, and Jann-Eve Stavesand, "Testing of Real-Time Criteria in ISO 26262 Related Projects - Maximizing Productivity Using a Certified COTS Test Automation Tool," SAE Technical Paper 2016-01-0139, 2016, doi:10.4271/2016-01-0139.
3. Jana Karina von Wedel and Paul Arndt, "Safe and Secure Development: Challenges and Opportunities," SAE Technical Paper 2018-01-0020, 2018, doi:10.4271/2018-01-0020.
4. Yifan Ye, Jian Zhao, Jian Wu, Bing Zhu, Yang Zhao, and Weiwen Deng, "Steering Control Based on the Yaw Rate and Projected Steering Wheel Angle in Evasion Maneuvers," SAE Technical Paper 2018-01-0030, 2018, doi:10.4271/2018-01-0030.
5. Ayush Goel and Somnath Sengupta, "Basic Autonomous Vehicle Controller Development through Modeling and Simulation," SAE Technical Paper 2018-01-0041, 2018, doi:10.4271/2018-01-0041.
6. David Parker, Yiannis Papadopoulos, Antoine Godof, and Laurent Saintis, "A Study of Automatic Allocation of Automotive Safety Requirements in Two Modes: Components and Failure Modes," SAE Technical Paper 2018-01-1076, 2018, doi:10.4271/2018-01-1076.
7. John Thomas, John Sgueglia, Dajiang Suo, Nancy Leveson, Mark Vernacchia, and Padma Sundaram, "An Integrated Approach to Requirements Development and Hazard Analysis," SAE Technical Paper 2015-01-0274, 2015, doi:10.4271/2015-01-0274.
8. Seth Placke, John Thomas, and Dajiang Suo, "Integration of Multiple Active Safety Systems using STPA," SAE Technical Paper 2015-01-0277, 2015, doi:10.4271/2015-01-0277.

9. Dajiang Suo, Sarra Yako, Mathew Boesch, and Kyle Post, "Integrating STPA into ISO 26262 Process for Requirement Development," SAE Technical Paper 2017-01-0058, 2017, doi:10.4271/2017-01-0058.
10. Junfeng Yang, Michael Ward, and Jahangir Akhtar, "The Development of Safety Cases for an Autonomous Vehicle: A Comparative Study on Different Methods", SAE Technical Paper 2017-01-2010, 2015, doi:10.4271/2017-01-2010.

In the following, we summarize the above papers in the context of the role of ISO 26262 in the development of AVs. After this introduction, the actual papers follow.

I.6.1 Discussion of the Role of ISO 26262 Papers

1. SuSanta Sarkar and James Forsmark, "System Engineering for Automated Software Update of Automotive Electronics," SAE Technical Paper 2018-01-0750, 2018, doi:10.4271/2018-01-0750.

   Automated software update of automotive electronics has evolved into a large and complex process; thus it stands to benefit from SE techniques. This paper discusses the component-oriented, manual process approach commonly used for software updates and proposes a SE approach. Some of the benefits of the proposed approach include a timely delivery of security-related updates, addition of features using software updates, and control of software update cost. The authors developed a suitable system architecture based on relevant system attribute including a model of a software update system. The developed model is compared with other models including measures to evaluate effectiveness.

2. Andreas Himmler, Klaus Lamberg, Tino Schulze, and Jann-Eve Stavesand, "Testing of Real-Time Criteria in ISO 26262 Related Projects - Maximizing Productivity Using a Certified COTS Test Automation Tool," SAE Technical Paper 2016-01-0139, 2016, doi:10.4271/2016- 01-0139.

   Some safety requirements for automated systems involve real-time constraints, and testing these constraints in software-intensive systems is challenging. This paper discusses the use of automated tools to maximize productivity for testing real-time constraints in ISO 26262-related projects. The authors argue that productivity increases can be achieved by improving the usability of software tools and decreasing the effort of qualifying the software tool for a safety-related project.

   Verification is particularly important for test automation tools that are used to run hardware-in-the-loop (HIL) tests of safety-related software automatically around the clock. This qualification of software tools requires advanced knowledge and effort. This problem can be solved if a tool is suitable for developing safety-related software. The paper describes signal-based testing is to define test descriptions as if sketching graphs on a sheet of paper with a plotter-like editor to intuitively describe stimulus signals and reference signals as a sequence of signal segments, thereby lowering the initial hurdle toward setting up an efficient test automation.

3. Jana Karina von Wedel and Paul Arndt, "Safe and Secure Development: Challenges and Opportunities," SAE Technical Paper 2018-01-0020, 2018, doi:10.4271/2018-01-0020.

   Advances in hardware and software have led to an interdependency between safety and security, and managing the ensuing requirements is challenging. Safety and security goals can conflict, safety mechanisms might be intentionally triggered by attackers to impact functionality negatively, or mechanisms can compete for limited resources like processing power or memory to name just some conflict potentials. But there is also the potential for synergies, both in the implementation and during the development. For example, both disciplines require mechanisms to check data integrity, are concerned with freedom from interference, and require architecture-based analyses. SAE J3061 introduces a process framework for cybersecurity development that is intentionally very similar to that of functional safety as defined in ISO 26262. The authors discuss how problems that can arise if functional safety and cybersecurity processes are not properly aligned and integrated into the overall development process. This paper proposes steps toward coordinated safety and security processes that can prevent such problems and show how such an approach can benefit from synergies between safety and security.

4. Yifan Ye, Jian Zhao, Jian Wu, Bing Zhu, Yang Zhao, and Weiwen Deng, "Steering Control Based on the Yaw Rate and Projected Steering Wheel Angle in Evasion Maneuvers," SAE Technical Paper 2018-01-0030, 2018, doi:10.4271/2018-01-0030.

   This paper is about evasion maneuvers of an ego vehicle by simultaneously controlling of steering and braking. The paper details a control system to jointly control the steering angle and differential braking signal to avoid collisions. The authors argue that steering assistance is needed to avoid collisions since this task is hard for drivers. Based on the control of the steering wheel angle by the optimal preview control algorithm, the differential braking control is obtained by using feedback from yaw rate and the projected steering wheel angle information to improve the accuracy of trajectory tracking and the stability of the ego vehicle in evasion maneuver. Simulation analysis based on the vehicle dynamic software (ASM) is conducted in typical maneuvers. Results show that the coordinated steering algorithm can further improve vehicle tracking accuracy and vehicles' stability when using the same collision avoidance trajectory under the limit of designed lateral acceleration.

5. Ayush Goel and Somnath Sengupta, "Basic Autonomous Vehicle Controller Development through Modeling and Simulation", SAE Technical Paper 2018-01-0041, 2018, doi:10.4271/2018-01-0041.

   This paper is about the development of control systems in a simulation environment using a test bench. A set of several coordinated PID controllers are designed that include functionalities of automatic cruise control (ACC) and AEB. The controller, based on practical data, is developed in simulation environment to primarily maintain safe distance from surrounding traffic objects while fulfilling requirements such as jerk levels, conditional braking, speed limits, etc. In this work, only a longitudinal controller is developed for low speeds (<30 km/h) and low throttle scenarios for which a four-wheel-based vehicle dynamics model is formulated excluding the nonlinear tire model. The captured prerecorded traffic video along with acquired throttle, braking, speed, and relative distance information is synced with the proposed controller simulation execution for correlations, wherein the acquired relative distance data is used as reference to run the simulations. The proposed practical data-based simulation test environment is successful in creating multiple test scenarios, and the developed longitudinal controller is able to satisfactorily autonomously control the vehicle in the desired manner.

6. David Parker, Yiannis Papadopoulos, Antoine Godof, and Laurent Saintis, "A Study of Automatic Allocation of Automotive Safety Requirements in Two Modes: Components and Failure Modes," SAE Technical Paper 2018-01-1076, 2018, doi:10.4271/2018-01-1076.

   The process of allocation of system safety requirements at the component level and throughout the entire development phase is complex and challenging. This paper describes an approach for safety requirements decomposition and subsequent allocation which has been implemented in the HiP-HOPS tool, and it leads to optimal economic decisions on component Automotive Safety Integrity Levels (ASILs). The paper first discusses automatic ASIL decomposition on an automotive example. Then the authors describe an experiment using two different modes of ASIL decomposition based on (a) the safety requirements of components or (b) individual failure modes of components. Protection against independent failure modes could, in theory, be achieved at different ASILs, and this will lead to reduced design costs. Although ISO 26262 does not explicitly support this option, the paper addresses the implications of this more refined decomposition on system costs but also on the performance of the decomposition process itself. Finally, the authors discuss the general need for increased automation of safety analysis in complex systems, especially autonomous systems where an infinity of possible operational states and configurations makes manual analysis infeasible.

7. John Thomas, John Sgueglia, Dajiang Suo, Nancy Leveson, Mark Vernacchia, and Padma Sundaram, "An Integrated Approach to Requirements Development and Hazard Analysis," SAE Technical Paper 2015-01-0274, 2015 doi:10.4271/2015-01-0274.

   The introduction of new safety-critical features using software-intensive systems presents a growing challenge to hazard analysis and requirements development. These systems are rich in feature content and can interact with other vehicle systems in complex ways, making the early development of proper requirements critical. Catching potential problems as early as possible is essential because the cost increases exponentially the longer problems remain undetected. However, in practice these problems are often subtle and can remain undetected until integration, testing, production, or even later, when the cost of fixing them is the highest. In this paper, a design methodology based on STPA is demonstrated to perform a hazard analysis in parallel with system and requirements development. The proposed model-based technique begins during early development when design uncertainty is highest and is refined iteratively as development progresses to drive the requirements and necessary design features. The technique is evaluated by applying it to a realistic but generic shift-by-wire design concept in two iterations with varying levels of detail. In addition, as the requirements and design evolve and change over time, the changes can be immediately analyzed for new hazards without repeating the entire analysis. The approach is also applicable even before requirements are developed, providing feedback when some of the most important decisions are being made instead of waiting for a finished design or model to begin an analysis. In this way, potential issues can be identified immediately and more efficiently, thereby reducing the need for future rework.

8. Seth Placke, John Thomas, and Dajiang Suo, "Integration of Multiple Active Safety Systems using STPA," SAE Technical Paper 2015-01-0277, 2015, doi:10.4271/2015-01-0277.

   As the pace of integration and complexity of new features rises, it is becoming increasingly difficult for system engineers to assess the impact of new additions on vehicle safety and performance. In response to this challenge, a new approach for analyzing multiple control systems as an extension to the STPA framework has been developed. The new approach meets the growing need of system engineers to analyze integrated control systems, which may or may not have been developed in a coordinated manner, and assess them for safety and performance. The new approach identifies unsafe combinations of control actions, from one or more control systems, that could lead to an accident. For example, independent controllers for auto hold, engine idle stop, and adaptive cruise control may interfere with each other in certain situations. This paper demonstrates a method to efficiently identify potential unsafe scenarios without requiring a complete enumeration or individual analysis of all possible scenarios. As a result, the approach is scalable to large systems with many controllers. In this paper, the method is demonstrated through a case study involving several driver assistance systems including advanced brake controls, advanced engine control, and advanced adaptive cruise control. Potential conflicts that would prohibit safe and successful operation are also efficiently identified, allowing engineers to develop suitable controls that prevent these conflicts.

9. Dajiang Suo, Sarra Yako, Mathew Boesch, and Kyle Post, "Integrating STPA into ISO 26262 Process for Requirement Development," SAE Technical Paper 2017-01-0058, 2017, doi:10.4271/2017-01-0058.

   Developing requirements for automotive electric/electronic systems is challenging, as those systems become increasingly software-intensive. Designs must account for unintended interactions among software features, combined with unforeseen environmental factors. In addition, engineers have to iteratively make architectural trade-offs and assign responsibilities to each component in the system to accommodate new safety requirements as they are revealed. STPA is a new technique for hazard analysis that considers hazards caused by unsafe interactions between components (including humans) as well as component failures and faults. STPA covers the safety analysis of system malfunctions as well as the SOTIF, in addition to functional safety. This paper describes a process map for integrating STPA into the functional safety process based on ISO 26262. Specifically, three steps in the process map are illustrated through a case study on an automotive system: (1) system assumptions and components from item definition are used to form the SE foundations for STPA; (2) UCAs identified and safety constraints created in STPA Step 1 are used to evaluate existing safety goals with ASIL ratings developed from HARA; and (3) causal scenarios and factors for UCAs identified in STPA Step 2 help engineers create functional safety requirements and make architectural decisions. In particular, the paper illustrates how STPA can help evaluate safety and other system-level goals with ASIL classifications from ISO 26262's recommended HARA. The meta-model can be also used to provide guidance on making architectural decisions in order to create functional safety requirements. To make the process map applicable to different functional safety processes adopted by OEMs, tool support is required. Guidelines on how to develop visualization tools based on the meta-model are given.

10. Junfeng Yang, Michael Ward, and Jahangir Akhtar, "The Development of Safety Cases for an Autonomous Vehicle: A Comparative Study on Different Methods", SAE Technical Paper 2017-01-2010, 2015, doi:10.4271/2017-01-2010.

    To achieve complete safety, a safety case providing guidance on the identification and classification of hazardous events and the minimization of these risks needs to be developed throughout the entire development lifecycle process of Connected and Automated Vehicles (CAVs). A comprehensible and valid safety case has to employ appropriate safety approaches complying with the automotive functional safety requirements in ISO 26262. This paper presents a comparative study of different safety approaches, in particular, FMEA method and goal structuring notation (GSN) method that have been employed to generate lists of hazardous events, safety goals, and functional safety requirements at the vehicle level. A case study on the safety case development of INSIGHT autonomous vehicle has been carried out using the aforementioned methods. This case study covers the safety argument of battery and charging system that supply the whole electric power for INSIGHT vehicle. The safety of this system has been assessed along with their potential for malfunction together with the layers of protection. The results and conclusions from case study analyses suggest the safety case of CAVs can be developed in a highly effective manner by employing a combined method of GSN and FMEA.

I.7 Conclusion

ISO 26262 is an important international standard useful for designing automotive systems with a high level of safety. Although ISO 26262 does not address some safety categories of AVs such as multi-agent safety and SOTIF, it is nevertheless still highly relevant to improve its safety. This is so because ISO 26262 uses a SE, risk-based, and lifecycle-based approaches to safety. SE is a powerful methodology particularly useful when dealing with large and complex systems such as AVs. Although SE has been used for many decades now in a wide variety of fields, we are seeing a renewed interest in its application to AVs, particularly MBSE using the SysML language. Another structure widely used in automotive and AVs is that of a feedback control system including other control systems based on the feedback loop, particularly *cascading control*. An example was discussed to implement most controllers in an AV as cascading controllers where the controller outputs of higher-level controllers become reference signals for lower-level controllers. An important activity when designing safe AVs is the identification of hazards, and two methods have been reviewed in this chapter, DFMEA and STPA. While the former is a generic technique that focuses on component failures, the latter focuses on control systems and SE structures and concepts. There is currently a great deal of interest in using both DFMEA and STPA for identifying hazards and designing mitigation solutions to reduce risk. This book collection includes ten papers covering the implications and use of ISO 26262 in the design of safe AVs.

References

1. Thrun, S. et al., "Stanley: The Robot That Won the DARPA Grand Challenge," *Journal of Field Robotics* 23, no. 9 (2006): 661–692.
2. Urmson, C. et al., "Autonomous Driving in Urban Environments: Boss and the Urban Challenge," *Journal of Field Robotics* 25, no. 8 (2008): 425–466.
3. Cheng, H., *Autonomous Intelligent Vehicles: Theory, Algorithms, and Implementation* (London, Springer, 2011).
4. Wang, F.-Y. et al., "IVS O5: New Developments and Research Trends for Intelligent Vehicles," *IEEE Intelligent Systems* 20, no. 4 (2005): 10–14.
5. Cheng, H. et al., "Interactive Road Situation Analysis for Driver Assistance and Safety Warning Systems: Framework and Algorithms," *IEEE Transactions on Intelligent Transportation Systems* 8, no. 1 (2007): 157–166.
6. International Organization for Standardization, "Road Vehicles—Functional Safety," ISO Standard 26262, 2011.
7. Koopman, P. and Wagner, M., "Autonomous Vehicle Safety: An Interdisciplinary Challenge," *IEEE Intelligent Transportation Systems Magazine* 9, no. 1 (2017): 90–96.
8. Avizienis, A. et al., "Basic Concepts and Taxonomy of Dependable and Secure Computing," *IEEE Transactions on Dependable and Secure Computing* 1, no. 1 (2004): 11–33.

9. Leveson, N.G., *Safeware: System Safety and Computers* (Addison-Wesley, 1995).
10. Wendorff, W., "Quantitative SOTIF Analysis for Highly Automated Driving Systems," *Safetronic 2017, Conference Proceedings*, Stuttgart, Germany, 2017.
11. Thomas, J., Sgueglia, J., Suo, D., Leveson, N. et al., "An Integrated Approach to Requirements Development and Hazard Analysis," SAE Technical Paper 2015-01-0274, 2015, doi: 10.4271/2015-01-0274.
12. Leveson, N.G., *Engineering a Safer World: Systems Thinking Applied to Safety* (MIT Press, 2012).
13. Young, W. and Leveson, N.G., "An Integrated Approach to Safety and Security Based on Systems Theory," *Communications of the Association for Computing Machinery (ACM)* 57, no. 2 (2014): 31–35.
14. Abdulkhaleq, A., Wagner, S., and Leveson, N., "A Comprehensive Safety Engineering Approach for Software-Intensive Systems Based on STPA," *3rd European STAMP Workshop, Conference Proceedings*, Amsterdam, The Netherlands, 2015.
15. Abdulkhaleq, A. et al., "A Systematic Approach Based on STPA for Developing a Dependable Architecture for Fully Automated Driving Vehicles," *4th European STAMP Workshop, Conference Proceedings*, Zurich, Switzerland, 2017.
16. Kopetz, H., *Real-Time Systems: Design Principles for Distributed Embedded Applications* (Kluwer Academic Publishers, 1997).
17. Koopman, P. and Wagner, M., "Toward a Framework for Highly Automated Vehicle Safety Validation," SAE Technical Paper 2018-01-1071, 2018, doi:10.4271/2018-01-1071.
18. International Electrotechnical Commission, "Functional Safety of Electrical/Electronic/Programmable Electronic Safety-Related Systems," IEC Standard 61508, 2010.
19. International Electrotechnical Commission, "Functional Safety—Safety Instrumented Systems for the Process Industry Sector," IEC Standard 61511, 2018.
20. Shalev-Shwartz, S., Shammah, S., and Shashua, A., "On a Formal Model of Safe and Scalable Self-Driving Cars," *Computing Research Repository (CoRR)*, arXiv:1708.06374 [cs.RO] (2017). [Online] http://arxiv.org/abs/1708.06374.
21. Gauerhof, L., Munk, P., and Burton, S., *Structuring Validation Targets of a Machine Learning Function Applied to Autonomous Driving*, Gallina, B. et al. (Eds.): *SAFECOMP 2018, LNCS 11093*, 2018, 45–58.
22. Feth, P., Adler, R., Fukuda, T., Ishigooka, T. et al., "*Multi-Aspect Safety Engineering for Highly Automated Driving Looking beyond Functional Safety and Established Standards and Methodologies*, Gallina, B. et al. (Eds.): *SAFECOMP 2018, LNCS 11093*, 2018, 59–72.
23. Pimentel, J. and Bastiaan, J., "Characterizing the Safety of Self-Driving Vehicles: A Fault Containment Protocol for Functionality Involving Vehicle Detection," *2018 IEEE International Conference on Vehicular Electronics and Safety (ICVES)*, Madrid, Spain, September 12-14, 2018.
24. Pimentel, J., Bastiaan, J., and Zadeh, M., "Numerical Evaluation of the Safety of Self-Driving Vehicles: Functionality Involving Vehicle Detection," *2018 IEEE International Conference on Vehicular Electronics and Safety (ICVES)*, Madrid, Spain, September 12-14, 2018.
25. Schorn, C., Guntoro, A., and Ascheid, G., *Efficient On-Line Error Detection and Mitigation for Deep Neural Network Accelerators*, Gallina, B. et al. (Eds.): *SAFECOMP 2018, LNCS 11093*, 2018, 205–219.

26. International Organization for Standardization, ISO/WD PAS 21448, "Road Vehicles—Safety of the Intended Functionality," ISO Working Draft, 2013.

27. Burton, S., Gauerhof, L., and Heinzemann, C., "Making the Case for Safety of Machine Learning in Highly Automated Driving," *International Conference on Computer Safety, Reliability, and Security (SAFECOMP), Conference Proceedings*, Trento, Italy, 2017.

28. Salay, R., Queiroz, R., and Czarnecki, K., "An Analysis of ISO 26262: Machine Learning and Safety in Automotive Software," SAE Technical Paper 2018-01-1075, 2015, doi:10.4271/2018-01-1075.

29. Geiger, A. et al., "Vision Meets Robotics: The KITTI Dataset," *International Journal of Robotics Research* 32, no. 11 (2013): 1231–1237.

30. X. Bi, Tan, B., Xu, Z., and Huang, L., "A New Method of Target Detection Based on Autonomous Radar and Camera Data Fusion," SAE Technical Paper 2017-01-1977, 2017, doi:10.4271/2017-01-1977.

31. Realpe, M., Vintimilla, B., and Vlacic, L., "Towards Fault Tolerant Perception for Autonomous Vehicles: Local Fusion," *7th IEEE International Conference on Robotics, Automation and Mechatronics (RAM), Conference Proceedings*, Angkor Wat, Cambodia, 2015.

CHAPTER 1

System Engineering for Automated Software Update of Automotive Electronics

SuSanta Sarkar and James Forsmark
General Motors LLC

In traditional automotive electronic design, software update has been a component oriented, manual process rather than a systematic designed in capability suitable for automation. In recent days as software content in vehicles grow, the need to update software in vehicles more frequently is becoming a necessity. Moreover, additional attributes for software updates, for example timely delivery of security related update for vehicles, desire to add features using software update, control cost of software updates, etc., requires a system engineered design rather than a component oriented approach. As the automobile domain utilizes various means of mobility (*Combustion Engine*, *Hybrid*, *Battery*, etc.) and various functional domains (*Infotainment, Safety*, *Mobility*, *Telematics*, *ADAS (Advance Driving Assist service)*, *Autonomous*, etc.), to control the overall cost of future software update for such a diverse environment, it is beneficial to introduce automation in the software update process. One way to facilitate introduction of automation is leveraging a systematic architecture. As the system that is used for updating software is also required to be updated, a stable interface design is very important.

The paper reviews the **state of practice** and **state of art** software update system architectures for the automotive domain. It also reviews the current software update architecture of the personal devices domain.

CITATION: Sarkar, S. and Forsmark, J., "System Engineering for Automated Software Update of Automotive Electronics," SAE Technical Paper 2018-01-0750, 2018, doi:10.4271/2018-01-0750.

Models of these architectures are developed to compare system architectures. To synthesize a suitable system architecture, relevant system attributes are established. Using the system attributes developed, a model of a software update system is synthesized. The developed model is compared with other models. In conclusion, measures to evaluate effectiveness are discussed.

Introduction

Historically, leading edge of automotive electronics trailed behind military and aerospace electronics. But, modern automotive systems are no longer less complex than the modern weapons or space systems. As features in automotive systems rely upon more and more electronics, use of software in the automotive systems is increasing at an exponential rate. For example, one of the Original Equipment Manufacturer (OEM) reported (2016) that the latest luxury model will have over 120 million lines of software code. Moreover, as automotive systems are becoming software intensive, the need to correct for errors, enhance existing features, and provide new software based features, is becoming more and more prevalent. OEM's have already begun (2015, 2016) adding driver assisted features using software update methods Automotive systems are going to use a lot more software and vehicle software will be required to be updated throughout its life cycle. Is traditional automotive electronic design and software architectural framework prepared for this challenge?

Automotive electronic systems are traditionally designed as embedded distributed systems with mostly broadcast data in local subnets. These systems consist of many electronic control units (ECUs) connected via various types of communication networks. Traditionally, automotive original equipment manufacturers have utilized suppliers to design and build these ECUs, and then assume responsibility of integrating them. In this framework, updating component software requires connecting an external tool and updating intended ECUs. This approach has the following issues persisting into the future:

- Each ECU's software update has been considered a standalone event from the supplier and performance perspective - the length of time it takes to update all software datasets in the intended vehicle ECUs is not a critical parameter for any one supplier.
- Simultaneous update of multiple ECUs is not designed in, but achieved as an artifact of individual ECU's update model, for example, communication network bandwidth required for simultaneous multiple ECU updates is not controlled by any one supplier.
- Software update dependencies among ECUs are not often explicit. Skilled operators at repair shops can mend implicit dependency using manual error detection methods and if necessary, subsequently installing additional software updates.
- Sometimes sensors or actuators connected to the ECU require manual manipulation following a software update.

If we consider that there are around 17 million vehicles being sold in USA alone, and each needing some form of software update every year, the burden of update using traditional framework and architecture - which is predominantly a manual process - is substantial. With an increase of software update events expected in the future, there is an opportunity to manage this using automation and better system design.

Moreover, recent focus on preventing unauthorized access of electronic systems in one hand, and introduction of ability to change software using automation on the other hand, introduces a major design challenge. The system that updates software is required to be completely secure and not add any vulnerability. A systematic approach rather than a component based approach has a better chance of achieving the objective.

The ability to correct errors using a software update allows for planning to correct a minor error after delivery of the product. From the contrarian perspective, in the automotive domain, the product is thoroughly validated before delivery and defects are resolved before product launch. Then does the software update feature provide value to the enterprise? As software becomes more complex, and features are used in a way not anticipated by the designer, sometimes an error will surface after the product is delivered. The ability to update software via automation enables the correction of these errors in a cost-effective manner. It also provides the ability to add more features in the future.

This paper is organized as follows: In the section, "*State of Automotive Practice and Models*", traditional and modern vehicle software update system are reviewed. In the following section, "*State of the Art and model*", software update architecture practiced in the personal device domain is described. Upon review, it becomes apparent that the existing software update system architectures used in the personal device domain are not entirely suitable for the current business driving forces and business environment for the automotive domain. In the section, "*Required System Attributes for Automated Software Update*", a list of system attributes is developed so that business drivers and current business conditions are satisfied. A model is synthesized so that all the system attributes developed earlier are realized in the model. In the section, "*Software Architectural Models required for Automated Software Update*", a software architecture is proposed so that the system that updates software is updatable. In *conclusion*, all of the models constructed are compared and guidance for evaluation is suggested.

FIGURE 1 Distributed electronic architecture of vehicles.

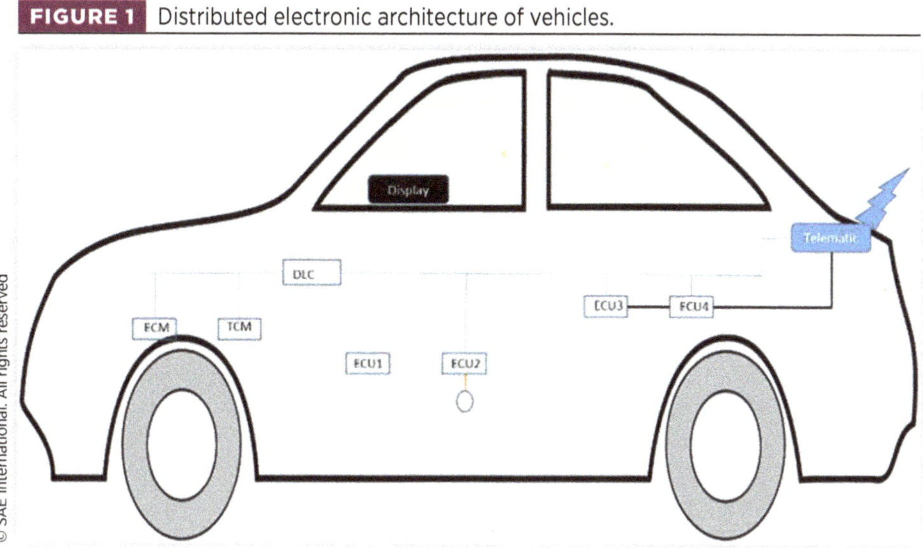

State of Automotive Practice and Models

Vehicle are predominantly updated either in the OEM assembly yard lot, immediately following production, or in the repair shop during servicing. As introduced earlier, the traditional method of automotive software update is performed by connecting an external tool to the vehicle's *Data Link Connector* (DLC) and following instructions in the external tool's user interface to achieve software update [4]. Time to update depends on the **size of the update, number of the modules to be updated, dependency of multiple module's order of install**, and number of **manual instructions to be followed**, if any. Since most of the ECUs inside the vehicle are connected to the DLC, most of the ECUs software can be updated using standard CAN bus diagnostics protocol. And, with a low volume of updates, a manual process does not pose an inordinate burden.

Earliest forms of automated software update have been used in the navigation system to update maps, and in the Telematics system to update the telematics software [1]. The primary driving force towards automation has been providing up-to-date maps without going through repair shop, and keeping up with changes in the cellular domain. In these cases, a standalone approach could still be used, as long as the target component to be updated is directly visible wirelessly from outside the vehicle, and the component to be updated does not interfere with the primary automotive features.

One of the hurdles of the wireless data download automation approach had been the cost of a wireless channel. One approach is to reduce the update payload size. Two competing approaches had been used: **Update Package Compression**, and **Update package reduction** using version difference. Update package reduction using a difference method works better if the expected updates are incremental and not structural. Depending on how compressible the complete image is, compression yields advantages. In some cases a combination of both techniques are employed to achieve a smaller payload. For larger changes, the difference method begins to take a longer install time. Certain images, such as Maps, are hard to compress.

More recently, one OEM considered software update as a significant business driver, and included automated capability to update vehicle software at the initial system design. Then a feature enhancement -involving brake, steering, and suspension - was delivered at the customer premise using only a software update delivered from a remote server using a wireless channel, [1]. In this case, the system had been designed from scratch, and had the advantage of a battery only powered vehicle's use cases - daily charging time, when the vehicle is not used. The system architecture realized for this case, provides an example for vehicles with other types of power train, the opportunity to control the burden of software update in future.

Models of Software Update Systems

State of the art systems can be modeled as "Traditional" or "Modern" as described below.

In "Traditional" system architecture, as shown in Figure 2, the vehicle is connected with a wired network. The programming tool, when manually connected and initiated, puts the target into programming mode and begins to deliver update package in segments. The update package has been obtained earlier from the back-office server based on valid updates available for a particular vehicle configuration. The vehicle is in a repair shop or assembly yard lot - no customer involvement is allowed.

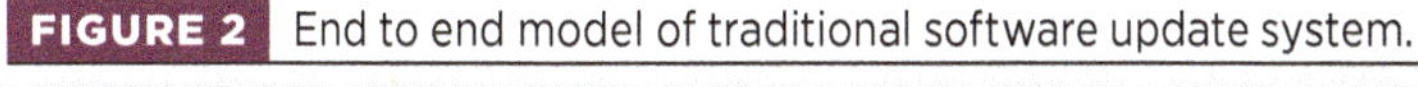

FIGURE 2 End to end model of traditional software update system.

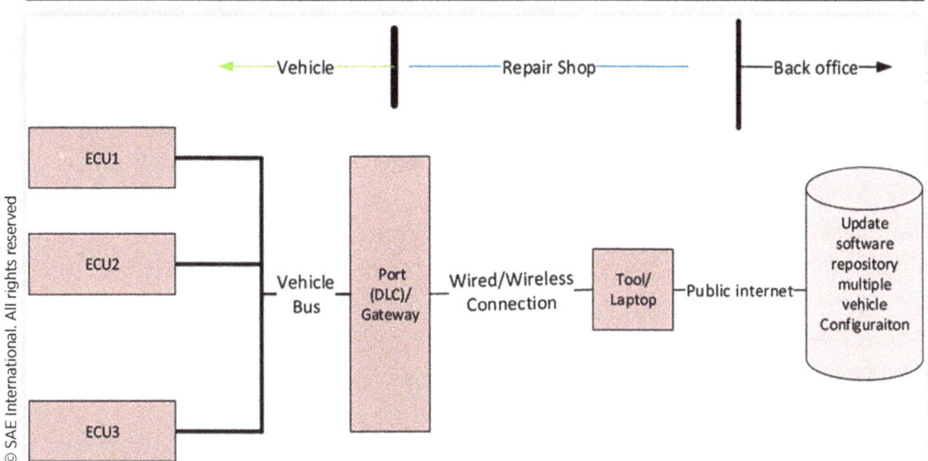

In "Modern" system architecture, the vehicle is connected with a wireless network. An update package exists, that is valid for a particular vehicle. It is downloaded and stored in the vehicle, or a storage device with update package is attached to vehicle, and then the on board programming tool updates one or more ECU's. In the case of Map update, the vehicle could be mobile. Minimal customer's involvement is required.

If mobility related features are being updated, the vehicle is rendered immobile. Minimal architectural considerations are required in the scenarios depicted in Figure 2, as: 1) automated update is limited to a few ECUs, 2) Interdependent ECU's are out of scope - exception [1], and, 3) how long the vehicle cannot be used is not a significant criterion, and 4) Scaling the system for the enterprise is a challenge. Storing of multiple configurations (various option content, model years, etc.), ready to be served is not a consideration. Moreover, the software update use case is viewed from the vehicle's component update perspective, without design consideration to reduce enterprise cost of software update.

FIGURE 3 End to end model of modern software update system.

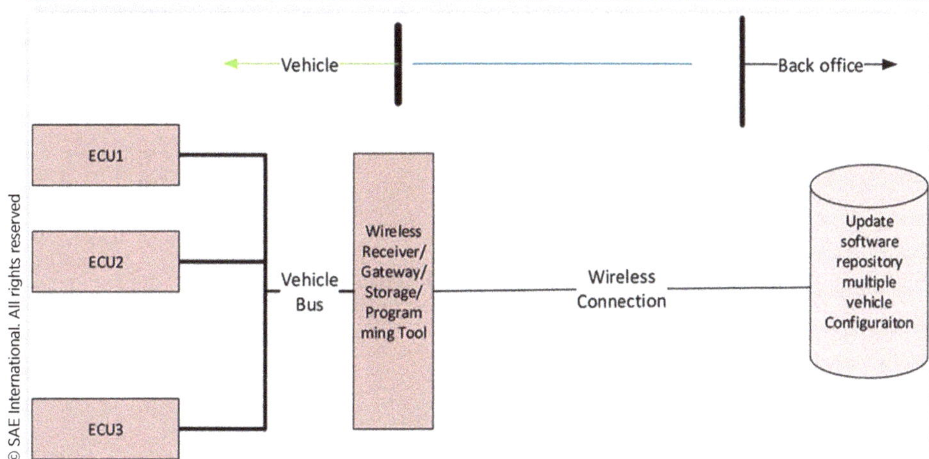

State of the Art and Model

Software update in the personal devices, such as: Cellular phones, Tablets, Lap Tops, and Desk tops had been accepted by customers as 'features' of such devices as this enabler is used to: offer new features, provision new devices, counter security related issues, etc. There are a few key architectural differentiators of these devices from automotive systems:

- First, these devices are viewed as a single endpoint. There could be many peripherals or many applications executing at the end point but the update package can be associated with a single end point.
- Second, these devices are connected to the internet network when powered, and usually the updates use bandwidth that is cost conscious. For example, the rule could be: use Wi-Fi (or wired internet connection) to update a tablet if it is connected to cellular and Wi-Fi.
- Third, as the consumer considers these as devices with software, prompting the customer to accept update, and agree to a temporary loss of usage had not raised dissatisfaction to an intolerable level. Even more recent usage of "Emergency Security Patches" for the consumer's benefit can be opted in.
- And if these devices are not being used, they are typically still powered. As a result, it is easy to find a time slot when the device is idle and suitable for an update.

Taking advantage of these aspects, most recent frame-works meant for personal devices are using the concept of synchronizing all endpoints with a master image - one of the aspects of the Mobile Device Management suite. When an update has to be delivered to multiple end points, the server starts a synchronization protocol that first, not only attempts to update the endpoint device as a whole, but also records the most up to date configuration of the end point to serve other business needs.

From an automotive perspective, the outcome is desirable but the context does not exactly fit traditional system architecture:

- Architecture to update an entire vehicle in one session is a challenge, as battery state of charge, and time to update are major constraints.
- Vehicle may not have sufficient memory to simultaneously store update packages for all components.

FIGURE 4 End to end model of state of the art software update system.

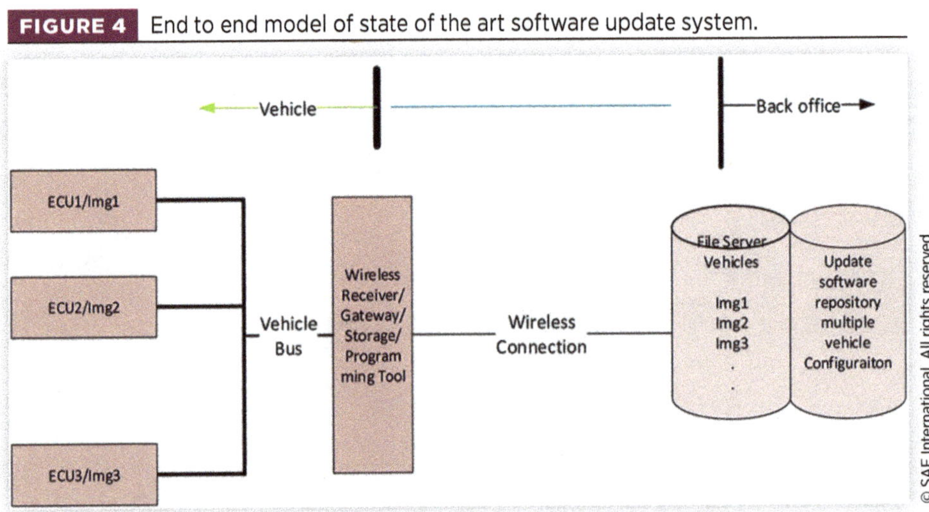

FIGURE 5 Personal device, synchronization software update model.

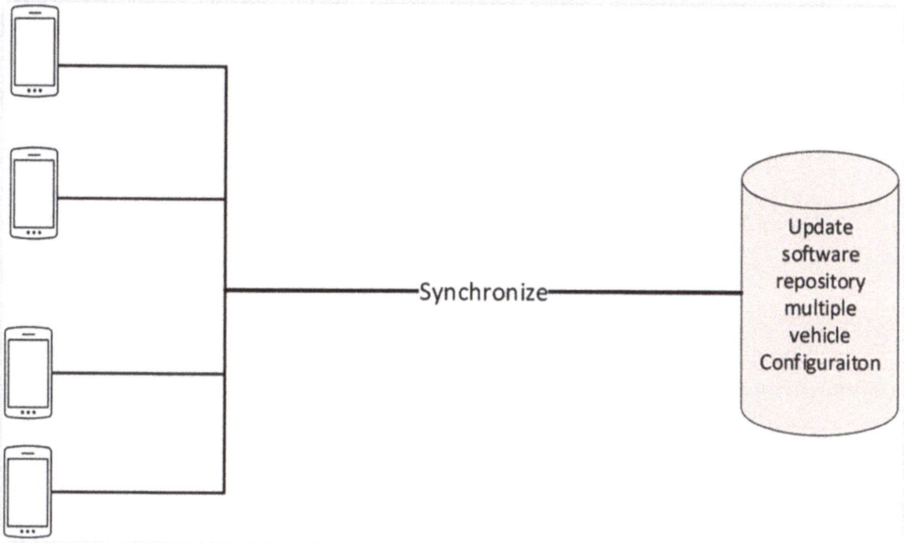

- Vehicle variability due to option content involves a large number of variants, and would require a large amount of storage at the back end.
- Vehicle life cycle, being an order of magnitude higher than personal devices, results in a large increase of the required number of generations of vehicles for software update to be supported.
- As vehicles are not perceived as a "Device with software", customer acceptance to update is a challenge, which may result in a long running update campaign.

Required System Attributes for Automated Software Updates

To arrive at an effective system design, relevant attributes are discussed in this section.

As pointed out earlier, an automotive electronic system is a distributed system of multiple electronic components connected by many buses. There are various attributes, described below, which contribute toward the automation of software update.

Safety

During an update of vehicle software, it is required to be safe. In the most stringent form, the vehicle is required to be immobile. Some preliminary part of the process could be accomplished when the vehicle is not immobile but the installation phase requires immobilization.

Connectivity

It is obvious that automatic update requires the ability of connecting the enterprise to the vehicle without manual intervention. Also, each component in the vehicle required to be connectable from an external wired connector(s) or wireless connection(s). However, this is not so straight forward. If security is considered - to be secure, minimal access is better. To prevent a malicious connection attempt, the connecting interface must provide **fire wall capability**. During software update, it is imperative that the target device is able to ascertain the commands and update package are coming from the intended source, and it is not tampered with on the way. Thus the architecture is required to **provide integrity and authentication** with various levels so that components cannot be compromised. Also, each component has the potential to be a custom designed electronic module, listed below are some challenges and potential solutions.

CAN CONNECTIVITY

The multi master, broadcast nature of CAN framework, makes it more challenging to implement authentication at the receiver's end. At a component level, each ECU can authenticate itself as a valid party to register to an in-vehicle master component - using PKI method. But, this adds much burden to smaller components. Traditionally **Seed/Key** algorithm had been used to this end. Newer architectures are using 'Key" based message authentication involving **specialized secure hardware** attachments.

Standard CAN protocol also constrains parallel software update to multiple components in the same subnet. **Special arbitration schemes involving interleaving** are required to be adopted in diagnostic mode to allow multiple components to receive segments of the update package from the "client", yet the "server" is able to send status to the "client" for the next segments, thus avoiding a timeout of the diagnostic session.

More recently, the idea of having redundant memory to provide a faster update and roll back capability is gaining ground. The redundant memory is updated in a variety of vehicle modes. However, CAN network bandwidth must be considered to ensure that programming communications does not interfere with functional communications. CAN buses with higher bandwidth are considerations for this case.

ETHERNET CONNECTIVITY

As automotive components are increasing in capability, usage of an Ethernet bus is being considered more often for higher performance. Not only does the advantage of a higher bandwidth leads to a faster update, but also utilities and designs from IT domain, e.g., security (integrity and authenticity), network transport, etc., are being reused to reduce the effort to complete a working design.

OTHER CONNECTIVITY

Many components integrate and/or control peripherals via other serial data bus such as LIN, SPI, etc. Typically, these peripherals can be configured by including their configuration package as a special partition of a parent component update. The parent component may either configure the peripheral every time it is updated or detect if peripheral is compatible with parent configuration and try to configure the peripheral.

Also, the architecture of some vehicles may result in a scenario where the update of a component interferes with the ability of an updater tool to communicate with other components that are downstream from this component. This requires the updater tool to implement a mechanism which can sequence the order of updates in an acceptable order which is compatible with the vehicle architecture.

Matching Bandwidth of Connection

One of the early lessons in the software update domain is, time to update a component depends more on the time it takes to transport the update package inside the automobile to the target than it takes to write into memory. The bandwidth of the data bus that connects a component to the external or wireless connector is required to match with memory size of the components.

Security

As the primary goal is to remove manual intervention, and move towards automation, it is absolute that no entity without proper authentication is able to change software in any component. Authenticated external connection from the vehicle as well as authenticated selective internal connection, and tamper proofing of the update packages are absolutely essential. Most designs include encryption for the remote connection, as it prevents tampering by man in the middle.

Local Storage

In a traditional ECU design, the programming tool stores the update package and it is delivered to the target ECU to be programmed one segment at a time, so there is no need to store the update package in the vehicle. In the automated design, vehicle connectivity to the remote source is realized using a wireless channel. In order to assure availability of the complete updated package, it is necessary to store it in the vehicle. So, it is required to have a storage device that can store update image of one or more ECUs.

In Vehicle Indication and Customer Interaction

The update process is required to inform the customer about changes to be made: How long it will take to update, reason for update, additional feature added by update, if any, etc. It also is required to communicate the status of the update, how much the update process has progressed, notify when the update is complete, etc. This attribute requires some means of indication to the customer.

The update process is also required to interact with the customer and guide through all the necessary steps the process will require. For example, parking in a safe and secure place, providing acceptance for the update, etc. This attribute requires some means of interaction with the customer.

High Success Rate

In the automated software update scenario, the focus is to reduce manual processes. Thus the success rate of the automated software update process is required to be very high. If an automated update is not successful, it will require manual intervention. Currently success rate is measured in terms of non-conformance measures like "x" per thousands, with a confidence factor. This measurement is required to be increased to one or two magnitude to avoid manual cost of rework.

Repeatability

Since, OEM's would like to update all components in vehicle, a defined process of preparing update packages, delivering them to the correct target, and recording installing success is required at the minimum. As an enterprise wide process, it is expected to work in the same way for all models and model years. Governance of this process includes monitoring the process, measuring success factors, and adjusting the process for improvement.

Battery State of Charge

The time the software update process takes and state of charge of the battery are closely related to each other. The vehicle shall be operable after the software update process. So, it is critical, that usage of battery during software update process is considered, and a proper bound is applied. A **battery state of charge measurement**, and indication feature is essential for this context.

Proposed Model of System Architecture

The proposed model of the system architecture is derived by extending the model of "Modern" version with the system attributes described in the section, "Required System Attributes for Automated Software Update". Following are the summarized enhancements.

- Overall safety is designed in such that it protects against known and perceived threats for all automated software update.
- Connectivity (Internet) using multiple wireless channels, in addition to wired interface are provisioned for.
- As multiple vehicle ECU's expected to be updated in one session, feasibility of bus bandwidth utilization relative to the state of battery charge of the vehicle is decided a priori.
- As automotive systems have long lifecycles, ability to update security system, are included in design.
- It is recognized that some ECU's could be too large to update using wireless transport, provision to update some ECU's with wired connection is included.
- In vehicle indication and interaction with the customer for impending software update is also explicitly identified.
- For automation to work, each ECU is required to have failure incidence measured per multiple thousands basis rather than the current practice of measuring incidences per thousand attempts.
- Each ECU is required to conform to some standard design and implementation details, so that software update works the same way for multiple vehicle models and multiple configuration within one vehicle model. It would be better yet, if the implementation can be controlled by one software source base, leading to higher chances for repeatability.
- As future vehicles will allow remote connection, logical interference relative to other connected features must be considered.

Software Architectural Models Required for Automated Software Update

Based on the system attributes described in the section, "Required System Attributes for Automated Software Update", software architecture and software elements can be organized considering various business drivers. However, as the lifecycle of vehicles are typically around a decade, it is essential that the software update system itself is also updatable. Systematic approaches to achieve this is to have a set of well-defined software elements that are decoupled as far as possible, and their interfaces deliberately designed not to change over time. This approach facilitates that changes in one software element does not ripple through the entire system. Software elements in this context could be mapped to one or more "objects" or "services". The architecture depicted in Figure 7 shows one way of such organization, and identifies important software based elements that are essential in realizing the system attributes identified earlier.

This architecture can be segmented in two broad categories, in-vehicle, and remote elements. Each software element is described in terms of its interface and capability at the application level. In-vehicle wireless service is an abstraction that implements brokering of remote connection and security related services between enterprise and vehicles for various modes of actual wireless types (cellular, Wi-Fi, etc.).

Following Are In-Vehicle Elements

BACK OFFICE INTERFACE (DOWNLOAD MANAGER)

Capability: The back office interface is one of the key components in the architecture. It listens for software update related commands from the enterprise, provides status to the enterprise, and when required, downloads files from designated servers. It is aware of the vehicle state including connectivity, maintains the vehicle state and software update state information, and interacts with stateless enterprise servers. It uses local storage to securely save downloaded file and enables/enforces security of the wireless connectivity.

Interfaces: Essential Interfaces are: a) Command/Status interface to Enterprise Servers, b) Vehicle Configuration information request/response, c) Remote File download

FIGURE 6 End to end model of proposed software update system.

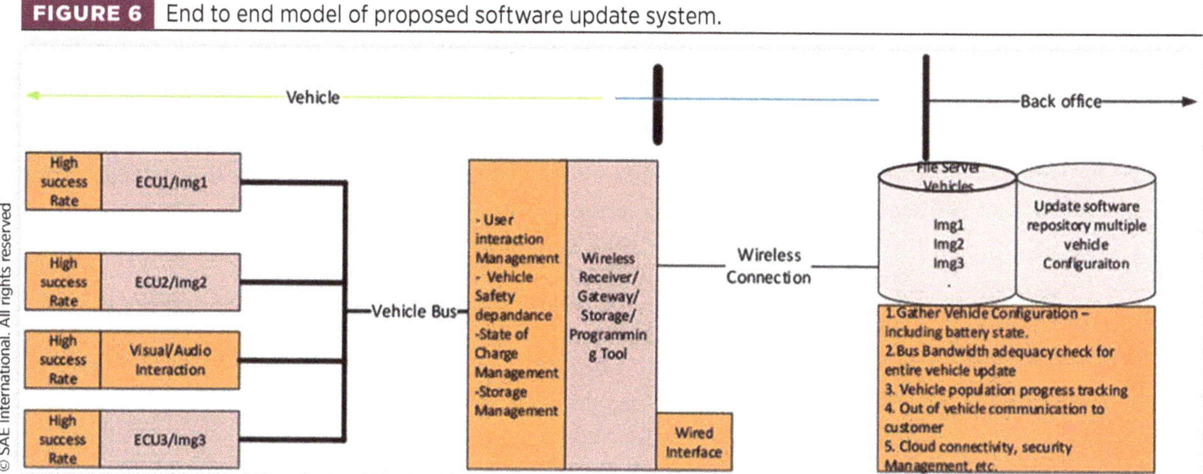

FIGURE 7 End to end software architecture for proposed automated software update process.

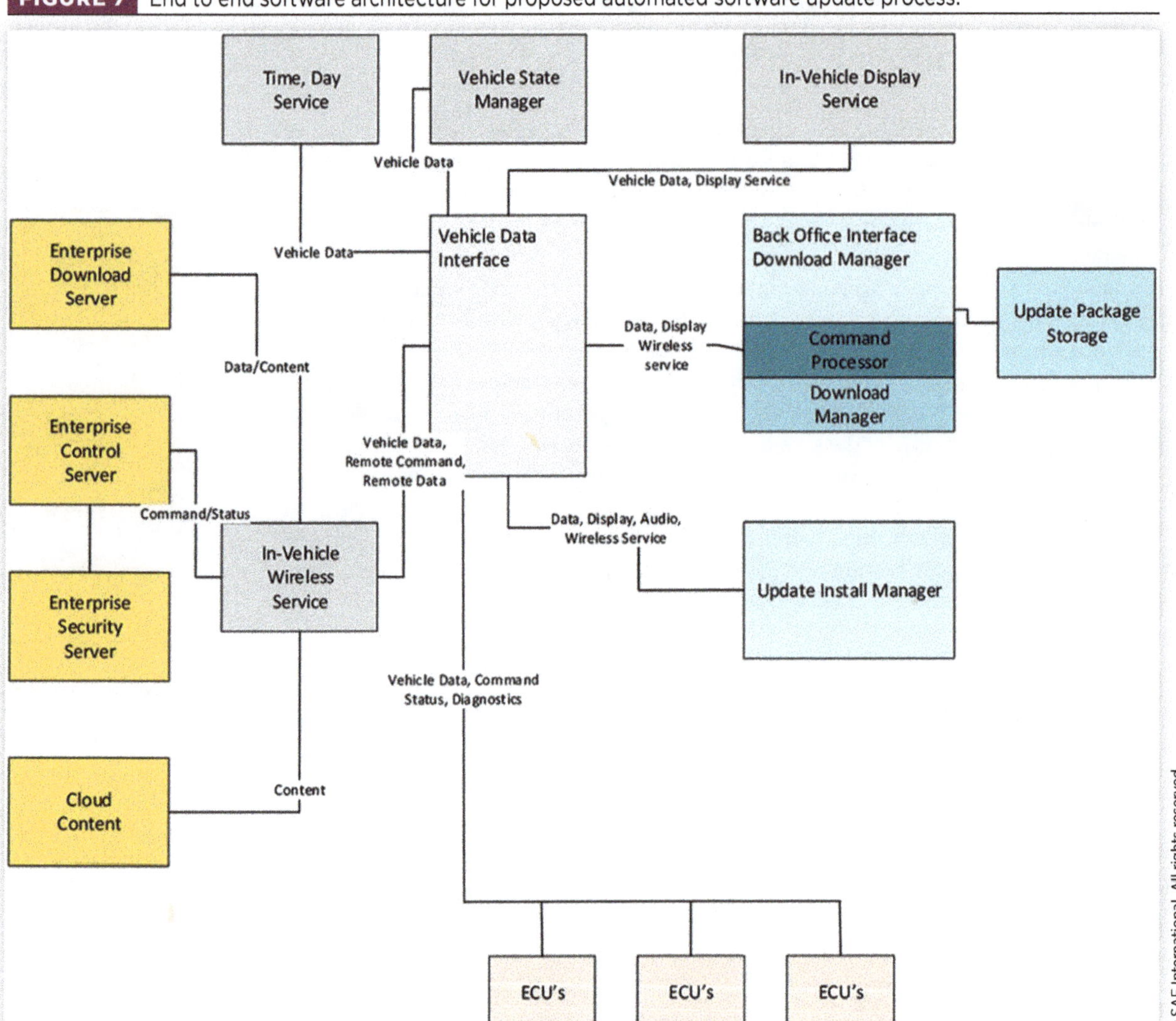

request/suspend/resume, d) Local file store/retrieve, e) wireless connectivity, f) Vehicle Data Services to be aware of vehicle state.

Security: At the minimum, authentication with the remote entity is required. Encryption is desired.

WIRELESS CONNECTIVITY SERVICE

Capability: Wireless Connectivity Service provides wireless connectivity capability to the vehicle - abstracts long or short range physical layer. It includes security to isolate the vehicle from un-authenticated connection. This also could be used to monitor network traffic for assigning the cost of connectivity.

Interfaces: Essential Interfaces are: a) Vehicle data services to be aware of the vehicle state, b) Long range and short range wireless physical layer abstraction, c) Enterprise Security server to initialize/renew certificates and other essentials.

Security: External to the vehicle interface at the minimum requires authentication with access point. For cellular service, the service subscription process includes authentication and authorization. For internal to the vehicle, some form of access control of vehicle data service is essential. This access control can be implemented using various means, for example, fire wall or authorization or authentication & encryption.

VEHICLE DATA INTERFACES

Capability: Vehicle Data interface holds the key to decoupling various components. It spans low speed (CAN) and high speed (MOST/ETHERNET/CAN FD/etc.) in vehicle networks. In the application level, techniques like publication/subscription without an intermediate broker model of dynamic connection can be used to isolate various feature elements. Also security related features such as authentication of sender, encrypted message, and prevention of replay attack should be configurable for specific data elements.

VEHICLE STATE MANAGER

Capability: User of the vehicle is allowed to change mobility, battery in use, connectivity, etc. related states of the vehicle, that are important for consideration towards automatically updating software in the vehicle. The traditional state manager is required to be augmented for automation by adding the ability to safely change state such that conditions required for the software update, such as vehicle not moving, sufficient state of charge, connectivity to internet, etc., can be achieved.

Interfaces: Ability to query the current state and request change of state, e.g., acquire connection, etc. are essential.

Security: For internal to vehicle, access control of the vehicle state manager is essential using authentication & authorization.

DISPLAY SERVICE

Capability: Initiation to download the update package, providing status of the download, initiation of the installation process, and providing status of the install - are best achieved using display and button/touch control. In order to decouple various kinds of user interface implementation, abstracted display service is essential.

Interfaces: Ability to instruct the customer to prepare the vehicle, and gain acceptance for automated update, and provide continuous status are the key interface elements.

Security: Access control based on state manager states and authorization, so that software update related interactions are not initiated either by rogue actors or inadvertently are required.

UPDATE INSTALL MANAGER

Capability: Update Install manager software element manages the various end to end transactions, and local services, such as providing status of vehicle configuration to the remote server, managing download of update packages, determining the vehicle state and initiating interaction with the user, and when all conditions are correct initiating update of the vehicle. Actual update process may use different kinds of libraries depending on the implementation.

Interfaces: It is key to have stable interfaces for remote server communication, download of update packages, and install services. As the install process is heavily implementation dependent, one of the challenging areas is to develop a stable and complete install interface that can use different transport, yet not have undue overhead. Careful considerations towards decoupling is necessary while developing this interface.

UPDATE PACKAGE STORE

Capability: As update Package storage service is expected to have many software update transactions in the life of the vehicle, stable and sufficient local storage of update package prior to beginning of installation is necessary. Also, storage management to keep room for the next update package, access control relative to other vehicle features using the storage service, maintaining integrity, and providing secure access are necessary sub-services. Remaining expected life relative to of read and write cycles for the memory must be included as self diagnostic.

Interfaces: Abstracted interface to the actual hardware storage, as well as exposing the storage service with a stable interface is necessary.

Following Are the Remote Elements

These software entities do not reside in the vehicle - these could be a single software element or part of a composite software entity.

ENTERPRISE CONTROL SERVER

Capability: As the remote server is aware of need for software update and communicates it to vehicles, services controlling the end to end system is more towards servers than vehicles. Also, in case of malfunction or a rogue element intruding the system, ability to disable part of the system resides in the remote server. It also could act as dispatcher to various other enterprise wide services and to the cloud.

Interfaces: Ability to initiate software update process, and gather status from initiation to completion are required. Ability to cancel an ongoing process for a set of vehicles is also desired.

Enterprise Download Server

Capability: Mostly for business reasons, automotive software for vehicle ECU's are stored in the OEM's premise. Even though repair shop service infrastructure requires that all of these are readily available, existing capacity could require expansion to accommodate automation. Moreover, as one of the transport hops is wireless, no undue server side delay can be accommodated.

Interfaces: Delivering file based on internet standard is recommended in this area, because of safety and security considerations.

CLOUD CONTENT ACCESSOR

Capability: As remote storage capacity utilization is dependent on software update usage, end to end design should be flexible to accommodate cloud hosting of storage service. Some of the commercial software components in the infotainment domain are better downloaded from the cloud.

Interfaces: Instead of developing proprietary software, adopting cloud service interfaces is prudent.

SECURITY SERVER

Capability: As stated earlier, security involves monitoring activities and, if required, updating to prevent access to rogue actors. It is essential to separate the security aspect of the remote service and maintain an in-house activity.

Interfaces: Ability to monitor key activities, ability to add/delete messages flowing across the system, govern certificate expiry and revocations are key elements.

Summary/Conclusions

Considering all system attributes listed earlier, capabilities of the "traditional", 'Modern", and "Proposed" Architectural model are summarized in Table 1. Label 'Point Design" is used if the attribute described in section 4.0 is not considered to the fullest extent,

TABLE 1 Summary of system attributes versus Architectural models

	Traditional	Modern	Proposed
Safety	Adequate	Adequate	System Designed
Connectivity	Wired connection.	Point Designed	System Designed
Total Internal Bus Bandwidth	Not in play.	Point Designed	System Designed
Security	Active hacking not anticipated	Point Designed	System Designed
Local Storage	Not in play.	Point Designed	System Designed
Remote Connection Management	Not in play.	Point Designed System Designed	
In Vehicle Indication and Customer Interaction	Not in play.	Point Designed	System Designed
High Success Rate	Not in play.	Point Designed	System Designed
Repeatability	Not in play.	Point Designed	System Designed
Battery State of Charge	Not in play.	Point Designed	System Designed
Software Architecture	Not in play.	Point Designed	System Designed

"System Designed" is used if the attribute is deliberately considered. The label "Not in Play" is used if the use cases at the point in time did not require this attribute.

As the "Proposed" architectural model is derived from synthesizing system attribute related capabilities form the "Modern" architectural model in practice, it satisfies the high level end to end system needs. But, how effective is the introduction of automation for software update in business operation?

One simplistic way to measure is to evaluate the cost per transaction of software update for automated and manual process. But this does not account for cost of upgrade to the entire system, and the operational cost of automation.

There are challenges in measuring the system wide impact of the proposed architecture. As common business practice, all new systems are evolved from the current one. Unless the enterprise transitions completely to the automated system, it is a subjective judgement how to assign weight to resources being leveraged by manual and automated process.

However, the projected volume of software update alone, directs towards automation, as the cost of manual operation in the future will be astronomical.

Contact Information

Dr. SuSanta P. Sarkar
General Motors LLC
susanta.p.sarkar@gm.com

James Forsmark
General Motors LLC
james.forsmark@gm.com

Acknowledgments

We acknowledge valuable comments provided by Kenneth P. Orlando, and insight regarding software installation process provide by Alan Wist.

Definitions/Abbreviations

ECU - Electronic Control unit
OTA - Over the air Programming
OEM - Original Equipment Manufacturer
ADAS - Advanced Driving Assist service
DLC - Data Link Connector
LIN - Local Interconnect Network
SPI - Serial Peripheral Interface
CAN - Controller Area Network
PKI - Public Key Infrastructure

References

1. Ahmed, M., “OTA Software Update Now Serving ECU’s for Engine Brake and Steering,” *Embedded Computer Design* (August 15, 2016).
2. Onuma, Y., Nozawa, M., and Terashima, Y., “Improved Software Updating for Automotive ECU’s: Code Compression,” *COMPSAC, IEEE*, 2016.
3. Nelson, G., “Over the Air Updates on Varied Path,” *Automotive News*, January 25, 2016.
4. Bulmus, A., Freiwald, A., and Wunderlich, C., “Over the Air Software Update Realization within Generic Modules with Microcontrollers Using External Serial FLASH,” SAE Technical Paper 2017-01-1613, 2017, doi:10.4271/2017-01-1613.

CHAPTER 2

Testing of Real-Time Criteria in ISO 26262 Related Projects—Maximizing Productivity Using a Certified COTS Test Automation Tool

Andreas Himmler, Klaus Lamberg, Tino Schulze, and Jann-Eve Stavesand
dSPACE GmbH

Increasing productivity along the development and verification process of safety-related projects is an important aspect in today's technological developments, which need to be ever more efficient. The increase of productivity can be achieved by improving the usability of software tools and decreasing the effort of qualifying the software tool for a safety-related project.

For safety-critical systems, the output of software tools has to be verified in order to ensure the tools' suitability for safety-relevant applications. Verification is particularly important for test automation tools that are used to run hardware-in-the-loop (HIL) tests of safety-related software automatically 24/7. This qualification of software tools requires advanced knowledge and effort. This problem can be solved if a tool is suitable for developing safety-related software. This paper explains how this can be achieved for a COTS test automation tool.

Maximizing the productivity of a software tool's use involves more than just considering the pre-qualification of the tool in accordance to safety standards. Enhancements in the tool's usability can also increase the productivity. Test automation tools usually let users develop tests by dragging blocks that define test steps. Signal-based tests are an advancement of this method, which is described in this paper. The aim of

CITATION: Himmler, A., Lamberg, K., Schulze, T., and Stavesand, J., "Testing of Real-Time Criteria in ISO 26262 Related Projects—Maximizing Productivity Using a Certified COTS Test Automation Tool," SAE Technical Paper 2016-01-0139, 2016, doi:10.4271/2016-01-0139.

signal-based testing is to define test descriptions as if sketching graphs on a sheet of paper with a plotter-like editor to intuitively describe stimulus signals and reference signals as a sequence of signal segments, thereby lowering the initial hurdle towards setting up an efficient test automation.

Introduction

Developers of safety-critical systems need to apply software tools to achieve their goals. This holds throughout applications in many industries, such as automotives, aviation, aerospace, nuclear technology, railways, medical fields, and the military. In these areas, it is crucial to verify the output of software tools in order to ensure their suitability for safety-relevant applications. Verification is especially important for tools that are used to automate certain development tasks, such as test automation tools. Test automation tools are used to increase the productivity required in today's development projects, as they make it possible to run tests on hardware-in-the-loop (HIL) test systems and assess the test results automatically 24/7.

Preparing HIL tests and assessing the test results need to be highly efficient in order to maximize the productivity of 24/7 test campaigns. Of course, this also applies to testing real-time criteria of safety-critical functions in automotive projects. In this context, the term 'functional safety' gains importance. The goal of functional safety is to ensure that highly complex overall products such as cars and especially their integrated E/E systems to not cause any danger for people and the traffic environment. Extensive software testing is necessary to reach this goal.

The overall goal of software testing is to detect unintended behavior in the software functionality, i.e., ensuring that the software is error-free and verifying that the required functionality is covered. Static and conventional test methods that are often performed without actually executing programs detect unintended behavior at the functional level, but cannot guarantee correct results under strict real-time conditions. Real-time testing is usually performed with hardware-in-the-loop test systems on which the software is executed under realistic environmental conditions. These systems can be used most efficiently when testing is fully automated. Hence, the HIL simulators are instrumented with a test automation tool that automates the complete test process, from executing test suites to generating the test result. Therefore, test automation and the relevant tool are essential for maximizing the effectiveness of the testing process. In addition, using HIL simulators to verify modern E/E systems is absolutely necessary because standards like ISO 26262 recommend this technology for implementing a useful testing process throughout the development cycle [7].

Users can improve the efficiency of a test automation by working with tools that are fit for purpose off-the-shelf to test safety-related systems. They also benefit from the vendor-provided guidance on how to use the software tools in safety-critical processes. In addition, the tools have to provide optimized usability to grant uncomplicated entry to test automation and test authoring, e.g., by letting users define test descriptions with a method like the one used for signal-based test descriptions. The aim of this method is to define test descriptions as if sketching graphs on a sheet of paper. In order to achieve this, a plotter-like editor is used to intuitively describe stimulus signals and reference signals for simulation variables. The documentation of the performed tests provides users a report that includes plots and additional parameter information.

This paper focuses on two aspects that increase the productivity of a commercial off-the-shelf (COTS) test automation tool: (a) minimizing the tool users' effort required for tool qualification and (b) maximizing the productivity for tool application. Productivity can be increased by using the new concept of signal-based test descriptions.

The paper is divided into four parts. The first two sections give a general introduction to the subject of tool qualification and to the automotive safety standard ISO 26262. The concept of pre-qualified software is introduced in the third section, along with the method for pre-qualifying software tools and how to use it. The last section describes the method of using signal-based test descriptions for a test automation tool. It also explains how this method increases the usability of test automation and how this is related to conducting tests under strict real-time conditions without expert knowledge.

Tool Qualification for Safety-Related Software

There are several safety standards that have been accepted and approved industry-wide, like IEC 61508, DO-178 for the aerospace industry, and ISO 26262 for the automotive industry. While 'IEC 61508 Functional safety of electrical/electronic/programmable electronic safety-related systems' is a basic functional safety standard that applies to all kinds of industries, the others are industry-specific. 'DO-178 Software Considerations in Airborne Systems and Equipment Certification' is a standard for the aviation industry and 'ISO 26262 Road vehicles - Functional safety' specifically applies to the automotive industry. All of these standards call for a tool certification or tool qualification process for software tools that are used to develop or test safety-critical systems. But each standard places a different amount of importance on tool certification.

The level of importance of tool certification differs across the standards. In general, users themselves have to learn to trust the software tools they use for testing and developing safety-related systems. But the tool vendors can increase user trust by providing a certain workflow for using their tools and a pre-qualification or certification according to the relevant standards.

This paper focuses on the automotive standard ISO 26262. Although it is considered a derivative of IEC 61508, it describes a significantly different method for qualifying tools for safety-related applications. ISO 26262 provides detailed assistance during the software qualification process [10], and thus for proving that a software tool is suitable for safety-related projects.

ISO 26262 Road Vehicles—Functional Safety

The increasing presence of embedded software in passenger cars has also resulted in increasing complexity, which is definitely an additional challenge when developing reliable systems. In addition, development projects need to comply with functional safety standards, because modern electronic control units (ECUs) control or interact with safety-related vehicle systems such as brakes and steering.

ISO 26262 is an emerging international safety standard. This standard is specifically designed for the automotive industry and is intended to be applied to safety-related systems that are based on E/E functionalities. ISO 26262 affects E/E systems that are installed in series production passenger cars up to 3.5 tons gross weight. For this, 'ISO 26262 Road vehicles - Functional safety' is an adaptation of IEC 61508, written to satisfy the needs specific to the application sector of E/E systems within road vehicles. This covers all activities during the development life cycle of systems composed of electrical, electronic, and software elements that provide safety-related functions. ISO 26262 is a risk-based safety standard. By qualitatively assessing operational situations that the risk of hazardous stages is related to, these safety measures are defined to avoid or control systematic failures to mitigate their effects.

The standard consists of 9 normative parts, from the concept phase over product development to the production and operation of the affected E/E systems, and a guideline for ISO 26262 as the 10th part.

Tool Classification and Qualification According to ISO 26262

ISO 26262 requires a detailed evaluation for each tool that is used in the verification and development of safety-critical systems. Each tool has to undergo software tool classification analysis (STCA), which is described and defined in ISO 26262. STCA consists of two steps. First, determining a tool confidence level (TCL) based on the tool impact (TI) and the tool error detection level (TD). Second, on the basis of the TCL and the automotive safety integration level (ASIL) of the development process in which the tool is used, additional qualifications steps and measures are required. ISO 26262 describes this as follows: "*Provide means for the qualification of the software tool when applicable, in order to create evidence that the software tool is suitable to be used to tailor the activities or tasks required by ISO 26262*" [9]

The tool impact (TI) indicates if and how the safety requirements of tool that the customer is considering to use are violated by any tool malfunction or erroneous tool output, and if this could violate the safety goals of the application. If a safety requirement cannot be violated, the tool impact level T1 is chosen. Otherwise the tool impact level is T2.

The tool error detection level (TD) is determined from the probability of preventing or detecting that the software tool is malfunctioning or producing erroneous output. The degree of confidence in this detection is divided into three tool error detection levels, ranging from TD1 to TD3. TD1 indicates the highest confidence, while TD3 indicates the lowest confidence. TD3 is chosen if there are no systematic verification measures in subsequent development phases or if malfunctions or erroneous outputs can only be detected randomly.

The tool confidence level (TCL) is determined on the basis of the tool impact and the tool error detection. Figure 1 visualizes the schema for determining the TCL.

Based on the TCL of a tool (i.e., the classification of a software tool) measures need to be taken to qualify the tool. In ISO 26262 the measures are defined in dependence of tool confidence level.

Table 1 lists the measures that are given for software tools classified as TCL3. For a TCL3-classified tool, ISO 26262 highly recommends that the tool is developed in accordance with safety standards for projects which are rated as ASIL D (marked by the two plus signs in Table 1). ISO 26262 describes that a relevant subset of requirements of the safety standards has to be selected for this investigation, because not every aspect of the standards are fully applicable [10]. In addition, the validation of the software has to demonstrate that the tool complies with its specific requirements; malfunctions and reactions to anomalous operating conditions have to be examined.

FIGURE 1 Schema for software tool classification according to ISO 26262.

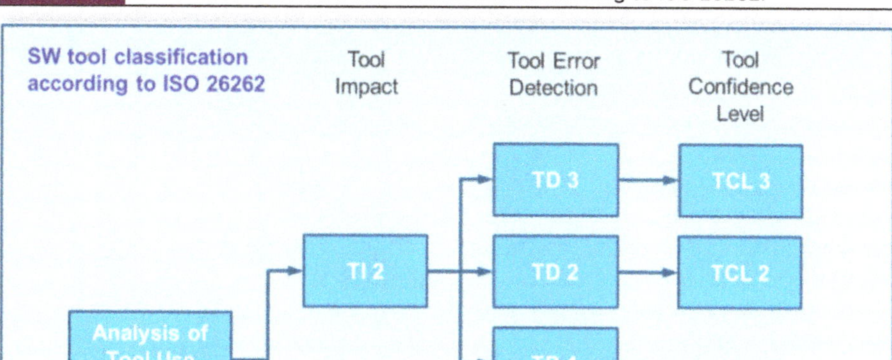

TABLE 1 Qualification of software tools that are classified as TCL3 [10] (+: recommended; ++: highly recommended)

Method	ASIL			
	A	B	C	D
Increased confidence from use	++	++	+	+
Evaluation of the development process	++	++	+	+
Validation of the software tool	+	+	++	++
Development in compliance with a safety standard	+	+	++	++

Relation to Product Liability

The classification and qualification of a software tool according to ISO 26262 described above is very important for developers of safety-related automotive systems. They should ensure that their development processes and also development tools are state-of-the-art, meaning that they are the highest level of development of a system at this particular time [4]. This ensures that a tool or a product is developed under the best possible technologies.

State-of-the-art development processes and tools are important for manufacturers of safety-related automotive systems since there is a close relationship to the manufacturer's liability for these products.

Under German law for example, manufacturers are generally liable for personal damage caused by the malfunction of a product [6]. But if the malfunction could not have been detected by the technological state of the art, the liability is excluded [11].

Thus, development in accordance to ISO 26262 is definitely a significant factor for manufacturers of safety-critical systems.

Impact on a Test Automation Tool

A test automation tool is applied to automatically access HIL test results, i.e., it automatically determines whether a test is passed or failed. It is very common to apply test automation tools for HIL tests that run 24/7.

Due to this kind of use, test automation tools are regarded to have the tool confidence level 3 (TCL3), since their malfunctions or erroneous outputs can only be detected randomly. The options users of such tools have to qualify their tool have already been described in the preceding subsection.

The actions that need to be taken to conduct such a qualification in accordance with ISO 26262 are a major initial hurdle for tool users. This is because advanced knowledge about ISO 26262 and the test automation tool under consideration is required. The required in-depth knowledge about ISO 26262 includes expert knowledge about tool qualification processes. In addition, tool qualification is most often a time-consuming task.

Users of the tools face the questions of how much expertise they want to acquire themselves and how much effort they can spend on this qualification process. These are crucial questions especially since development projects are nowadays always under very strict constraints with regard to time and budget.

Certified COTS Test Automation Tools

The problem of investing additional effort for qualifying a software tool in accordance with ISO 26262 can be solved if the tool under consideration is fit for purpose for developing safety-related software according to IEC 61508 and ISO 26262, and if the tool is prequalified for all ASILs according to ISO 26262. This means that the software tool is ready to develop or test safety-critical systems 'off-the-shelf'. Tool vendors qualify their tools according to ISO 26262 and thus relieve their customers of this burden. The customers can concentrate on incorporating the software tool into their verification and development process without spending additional effort on certifying or qualifying this tool. The effort of analyzing the tool's use cases and the impact on the verification process of safety-critical systems, which is described in the previous sections, passes over to the vendor of the software tool.

Nevertheless, the users of the software tool themselves need to ensure that the tool is working properly in their working environment. Although this can only be done by the users, they can be supported by the tool vendors. The vendors can provide guidance for how to integrate the software tool into this working environment.

This section describes how a COTS tool can be qualified in accordance to a standard like ISO 26262. The test automation tool AutomationDesk will be used as an example to show what qualification means in this case, and who is capable of performing such a qualification.

Test Automation Tool—Certification According to ISO 26262

A test automation tool is mainly intended for test development, library development, test execution, test result generation, and report generation.

There are two main challenges for the users of a TCL3-classified test automation tool, as already described:

1. Qualifying the software tool in accordance to ISO 26262
2. Integrating the software tool into their often unique verification process

The specific example of AutomationDesk shows how to address these challenges by pre-qualifying the tool in accordance to ISO 26262 and by giving the users guidance in integrating the software tool into the verification process and providing the features that are needed for such a process. The guidance is given by a reference workflow.

Qualification in accordance to ISO 26262 can be achieved through certification by an independent authority. A certification is a confirmation of an object's characteristics. The certification in accordance to standards, so the proof that the object is capable of fulfilling a certain standard can be provided by independent organizations or authorities like the German authority TÜV. TÜV works on validating the safety of nearly all kinds of products. Six TÜV companies share the same trademark but are independent companies and competitors, like TÜV Nord, TÜV Rheinland, and TÜV SÜD. TÜV SÜD, for example, works as an independent consultant that examines the qualification of software tools in accordance to ISO 26262 and has the authority to certify such a product in accordance to standards like ISO 26262 and IEC 61508.

In the case of AutomationDesk, TÜV SÜD assessed several aspects of AutomationDesk's development process to verify its capability for use in developing safety-critical applications. In particular, TÜV SÜD focused on the requirement, change and release management in AutomationDesk's development process. As mentioned above, verifying a tool classified as TCL3 involves not only examining the tool's impact in productive use, but also evaluating the tool's whole development process. In addition, this also entails assessing the customer information process and bug reporting (defect management, known problems).

As a result of this assessment, TÜV SÜD has certified that the latest version of AutomationDesk is suitable for testing safety-related systems according to ISO 26262 and IEC 61508. The TÜV SÜD certificate confirms AutomationDesk's suitability for developing and testing safety-related systems in automotives, commercial vehicles, aerospace and other industries. [3]

Users do not have to qualify such pre-qualified automation software again, but they are still responsible for integrating the tool into the verification process. Tool vendors could support the users by providing a reference workflow for the integration process and the required features. This is described in the following section.

Integrating a Software Tool into the Verification Process—Reference Workflow

Now the pre-qualified software tool has to be integrated into the users' verification process. While the tool users have to perform this task themselves, tool vendors could provide a reference workflow for the tool and the required features to guide users through this process. The software tool has to be properly integrated into the software environment of the user by careful installation and commissioning.

In the case of AutomationDesk, this guidance is provided by the AutomationDesk Safety Manual [1]. The manual explains how to use AutomationDesk for developing safety-related applications according to ISO 26262 [9] and IEC 61508 [8], how to integrate AutomationDesk into the software- and test environment, and how to handle any errors and known issues. Therefore, the safety manual mainly answers three questions with respect to AutomationDesk:

1. How can a test automation tool for HIL tests of safety-related systems be used effectively and systematically in order to ensure that all the required functionalities of the system under test can be tested and finally released?

FIGURE 2 Reference workflow for test automation.

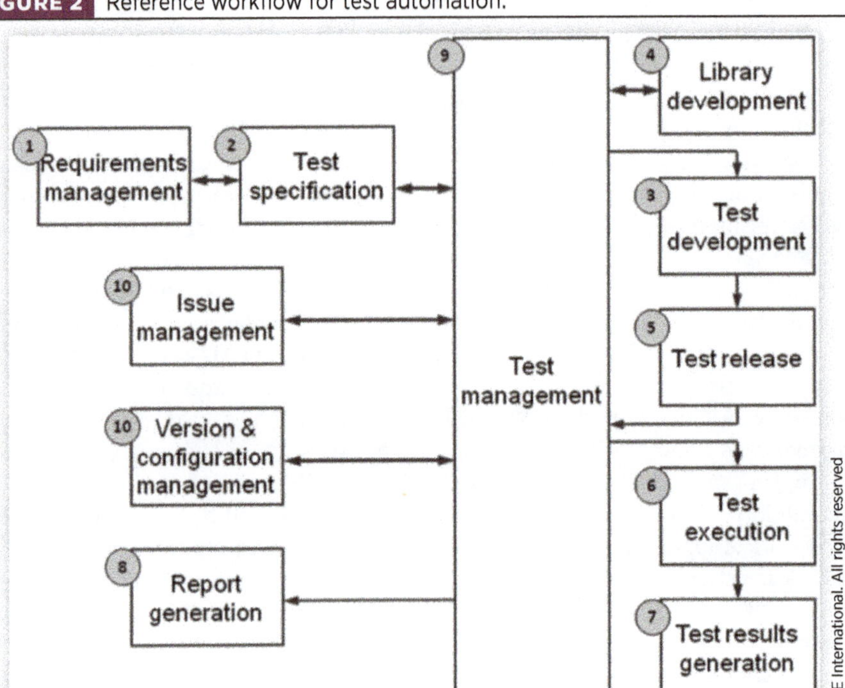

2. How can a test automation tool's suitability for testing safety-related E/E systems be verified in terms of confidence and appropriateness?
3. How can the requirements of the respective safety standard regarding the verification of safety-related systems and the suitability of software tools be met? [13]

AutomationDesk provides a central test project management capability for managing test sequences, libraries, test data, and test results and for generating test reports in different formats (HTML, PDF), and thus also provides basic test management capabilities [13]. This makes it possible to describe a reference workflow for the test automation process, including test management (Figure 2). The reference workflow and the associated test process could be integrated into the overall test process and into the software environment of the user.

With the TÜV SÜD certificate and the guidance for the verification process, it is possible for users to implement a testing process of safety-related systems more efficiently than by performing every single step on their own. This guidance for integrating AutomationDesk into a customer's verification process lowers the hurdle towards setting up an efficient test automation process significantly [1].

Signal-Based Testing Description for Testing Real-Time Criteria

The previous sections describe how the pre-qualification and certification of COTS software tools in accordance to standards enhances productivity, while decreasing the effort to integrate a COTS software tool into a development environment. Another factor that increases productivity is how usable the software tools under consideration are for developing safety-related systems. By improving the usability of a software tool, its initial hurdle towards productive use is shortened. This especially holds for test automation tools because test automation is an important factor in benefitting from all the capabilities of real-time platform such as a HIL simulator (e.g., through continuous operation). The possibility of setting up test automation without expert knowledge can be a leap forward in efficiency, which is why users demand test authoring and test automation tools that provide a graphical user interface. This is especially true in comparison to script-based concepts. Graphical user interfaces let users develop tests graphically by simply dragging blocks that define test steps. This enables users to create and edit test routines in a graphical format without any expert knowledge. In this case, usability means that the initial effort for setting up test automation could be minimized. This includes the test description and also the test evaluation because the purpose of test routines is to examine functionalities. Configuring the right evaluation for the test sequence involves at least as much effort as setting up the test sequence itself. In many cases it is necessary to evaluate the tests under strict real-time conditions to increase the test reproducibility (e.g., starting tests at the same time or under the same conditions) and traceability of the tests (e.g., mapping test results to events or timings). This also scales up the effort of implementing a suitable test evaluation and an associated test report.

Therefore the deterministic testing of real-time criteria also has to be taken into account to benefit from a real-time platform like a HIL simulator. The test automation tool has to combine aspects of usability and offer features like real-time testing to increase productivity throughout the verification process.

To describe the aspects of usability and testing real-time criteria, this paper again uses the example of AutomationDesk. AutomationDesk provides graphical test development and gives the users access to simulation platforms and tools from different vendors from within the automation sequences via the ASAM XIL API standard. For example, users can access HIL simulators to perform real-time stimulus, including with recorded measurement data, and they can access calibration tools [14]. AutomationDesk provides libraries, which enables the users to build test sequences from scratch without advanced knowledge of the tool itself. It is possible to create an initial test automation by using only libraries, but users can also customize the test sequences to various degrees. For example, they can use their own Python scripts. This lets users include the test automation tool into a verification process without having expert know-how.

The gate to this test automation programming is the Sequence Builder (Figure 3). This is a graphical editor which is control-flow-oriented and represents a UML-based description language [2]. After users edit the test sequence, the test runs on the test system by accessing the real-time model variables and parameters, e.g., via the ASAM XIL Model Access port. The test automation tool, which runs on the host PC of the HIL simulator, operates the test system by accessing the simulation model or electrical failure simulation.

As mentioned above, graphical test automation authoring is one important feature of test automation tools because it helps users work efficiently on the test automation. Signal-based test descriptions take this one step further. In contrast to block-based programming, users can describe tests intuitively as if sketching on a piece of paper. They can define real-time stimulation and evaluation rules for captured signals and evaluation functions, which lets them perform time- and amplitude-tolerant signal evaluations. Segment-wise stimulus or evaluation rules can be combined. Conditions for a transition between signal segments can be defined as real-time observers to check preconditions for subsequent segments, or as dynamic wait functions for synchronizing segments or for passed/failed evaluations (Figure 4). The documentation of the performed

FIGURE 3 AutomationDesk's sequence builder.

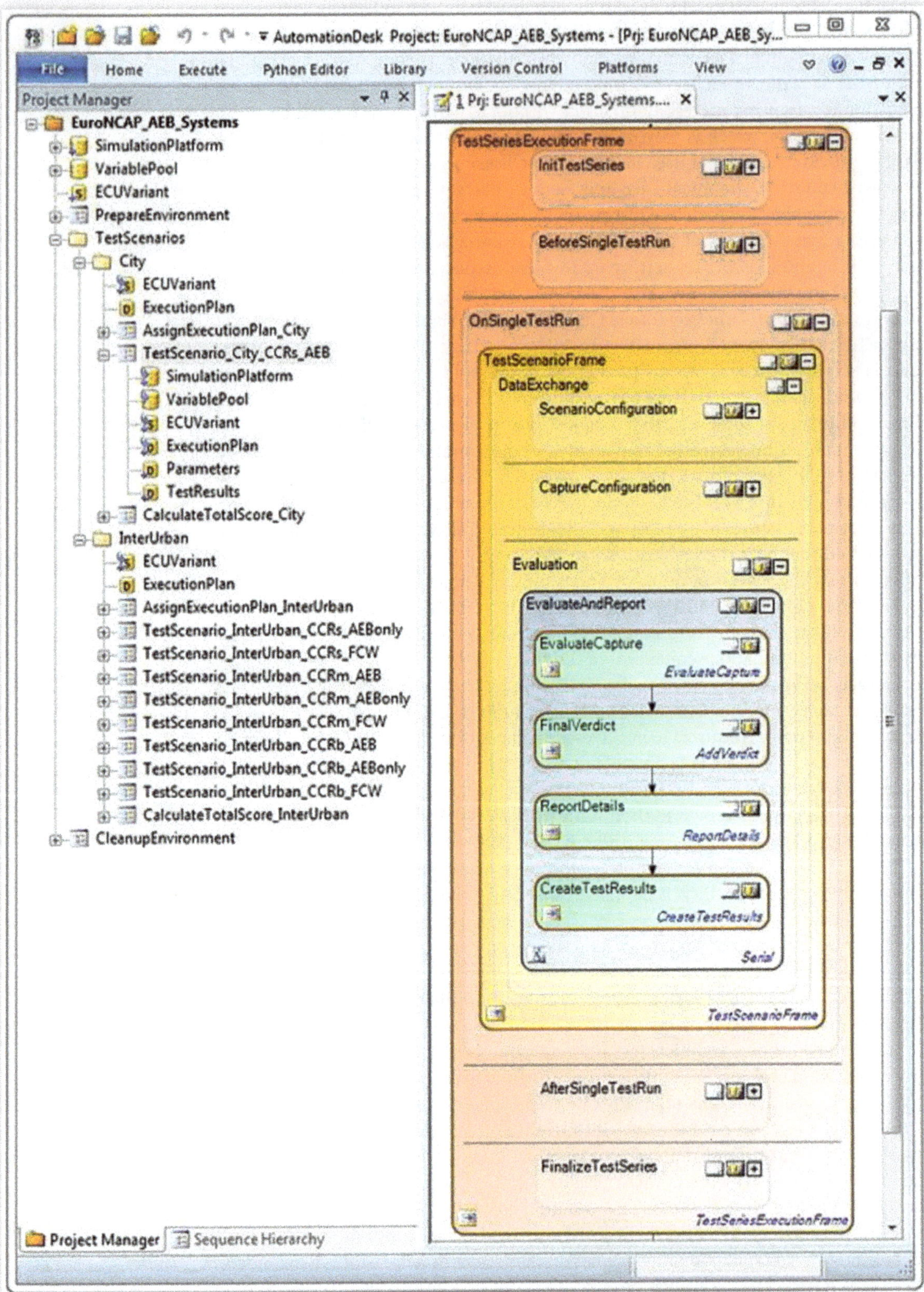

FIGURE 4 Signal description via defining signal segments.

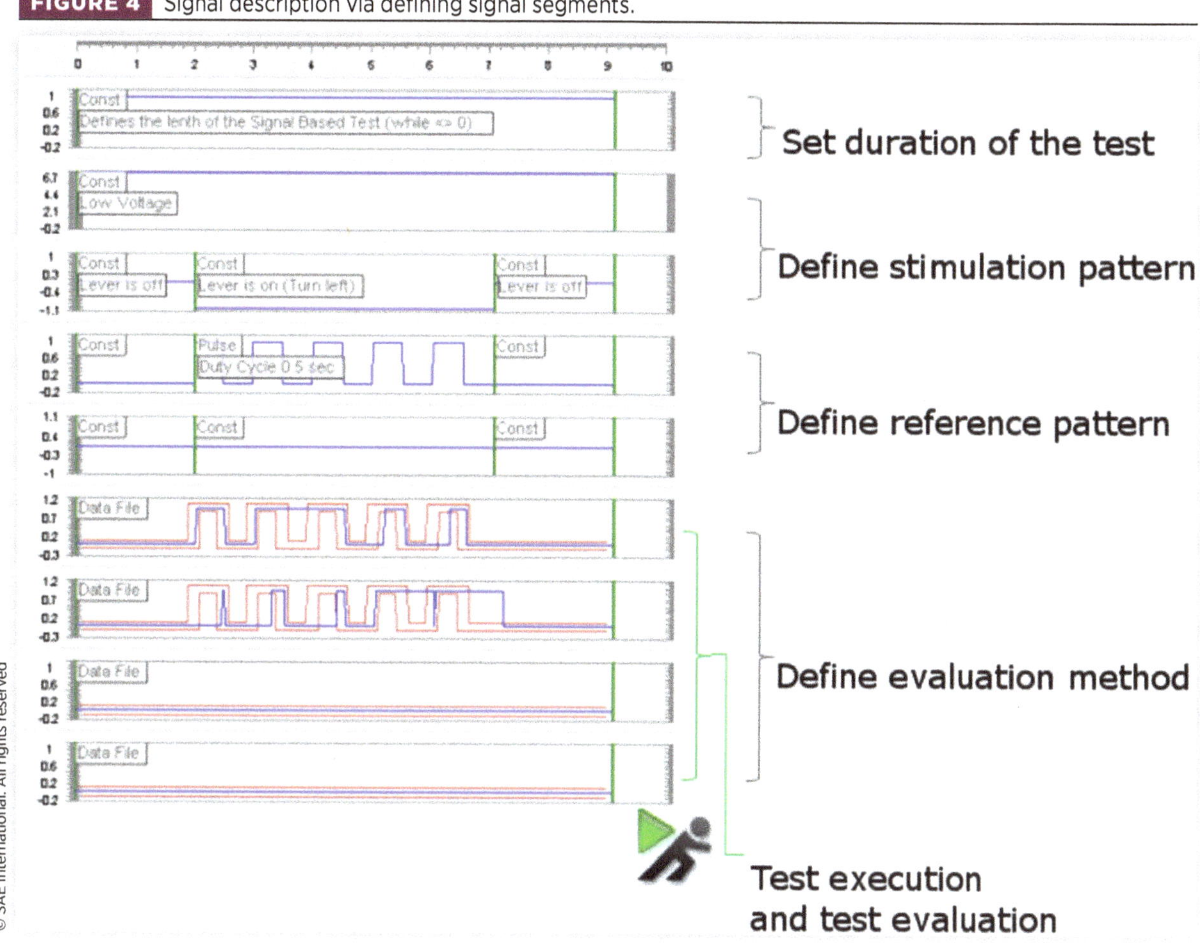

tests provides users a report including plots and additional parameter information. An ordered collection of signal segments can be defined for stimulus signals. These signals can either be defined synthetically or replay data from measurements. The evaluation of the monitored signals is also signal-based. For each monitored signal, just one evaluation rule needs to be defined for an entire signal. The evaluation rule consists of a tolerance tube that is defined by value- and time-based tolerances. These can be built by a sequence of segments in time (Figure 5).

This expands the testing capabilities by real-time criteria. The stimulus and reference signals are set on the real-time platform itself by executing Python scripts directly on the real-time core, decoupling the test process from influences of the host PC which does not have any real-time operating system. This makes timings more precise and the reproducibility and traceability of tests becomes much more efficient because each test execution follows precisely the same timing conditions.

The difference between stimulation of the real-time model from the host PC and the execution of real-time tests directly on the real-time platform is that signal-based tests do not require constant access from the PC to the real-time platform. The test execution is decoupled from the PC and therefore decoupled from influences of the not real-time capable operation system of the host PC (Figure 6).

The evaluations based on the conditions from the signal-based testing processes are automatically analyzed and visualized in a report (Figure 7). As mentioned above, the

FIGURE 5 Evaluation methods.

FIGURE 6 Real-time testing.

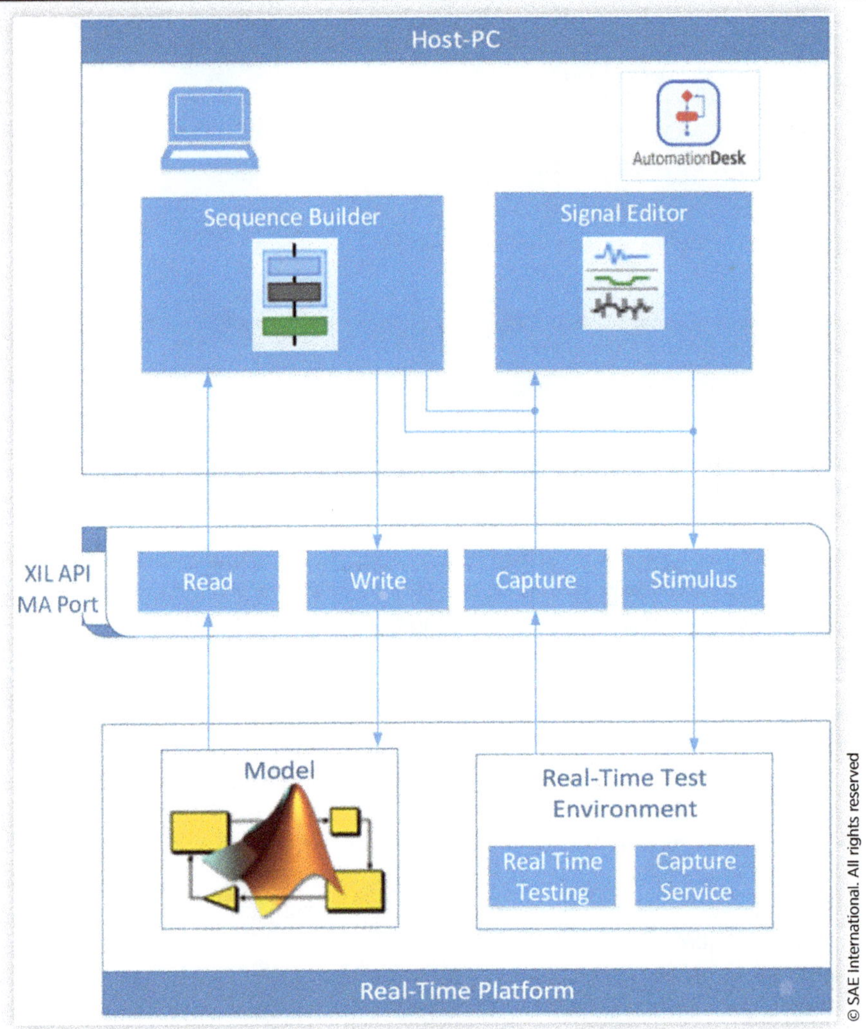

FIGURE 7 Automatically generated test result based on testing conditions.

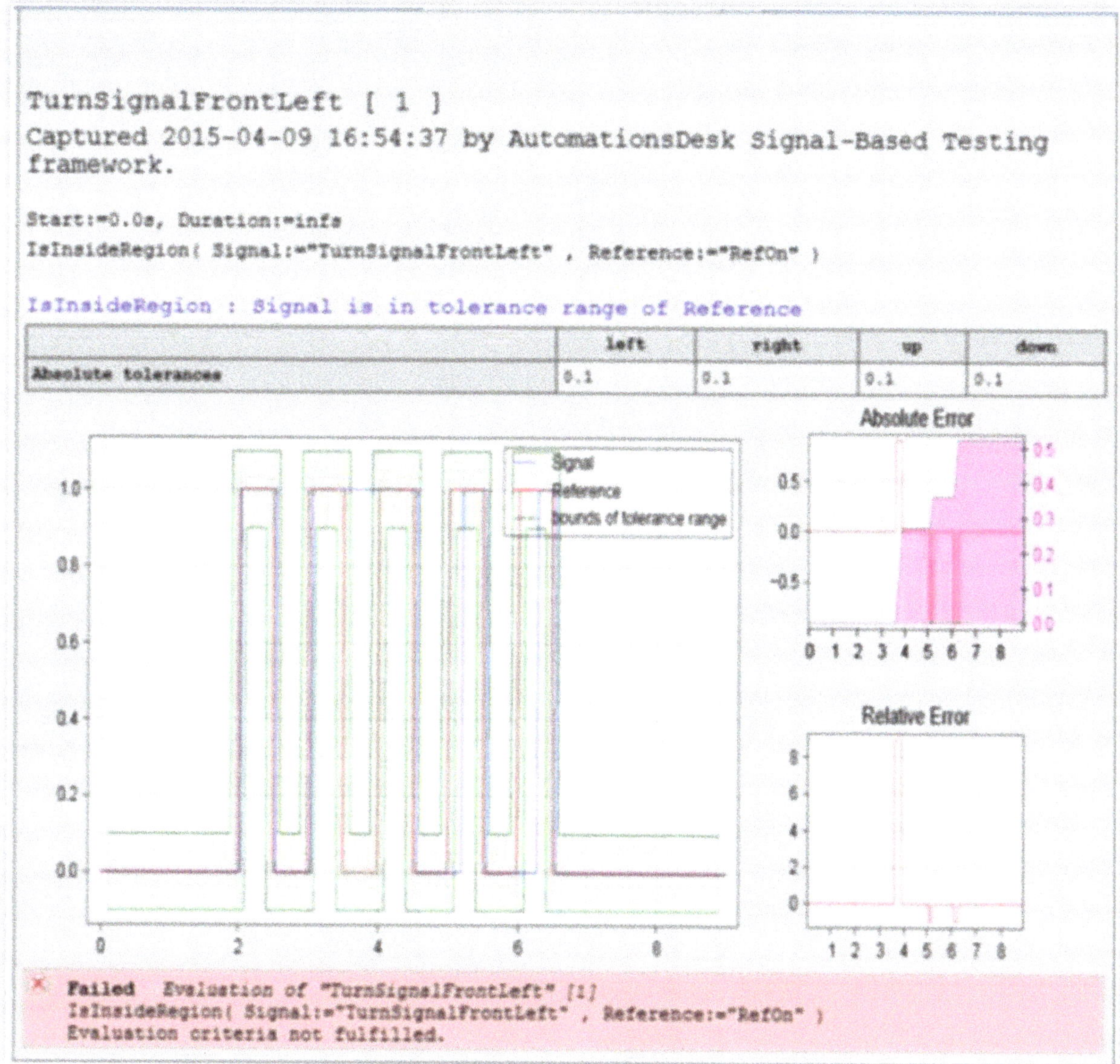

evaluation is also described by signals. The evaluation signals are also visualized in the test report and give a direct overview of the timings and the conditions which are used for evaluation. This gives users the possibility to examine the test under strict real-time conditions.

There are different technical methods for stimulating tests directly on the real-time system, e.g., by using MATLAB S-functions in the real-time model. But there are several disadvantages to these methods, such as the need for rebuilding the real-time application due to changes in the test procedure. It is not necessary to compile or build the real-time application again because AutomationDesk's signal-based method uses scripts. The users can edit and change the test description independently of the real-time application and the real-time execution of the models on the test system. AutomationDesk provides a graphical user interface for straightforward access to the functionalities of the tool. With the Sequence Builder in AutomationDesk, it is also possible to call Python scripts directly on the real-time platform. This could be another method for testing real-time criteria, but for beginners in Python programming, signal-based testing could significantly increase productivity.

Using Signal-Based Test Descriptions—Euro NCAP Autonomous Emergency Braking (AEB)

One use case for signal-based test descriptions is the deterministic maneuver control for European New Car Assessment Programme (Euro NCAP) test catalogs and the evaluation of the test catalogs under real-time criteria. Again using the example of AutomationDesk, this paper shows how users can benefit from using the signal-based method in this concrete use case. This use case will be described below after a shot summary of the Euro NCAP test campaigns.

AutomationDesk offers users a free-of-charge Euro NCAP test suite. This test suite offers all the tool configurations required to verify an ECU function according to the Euro NCAP testing scenarios. The Euro NCAP test suites benefit from the functionalities that are available only for signal-based testing, such as the intuitive test description, the deterministic test execution, and the flexible evolution according to tolerance definitions.

Euro NCAP uses crash tests and other forms of investigation to assess the safety of cars according to a five-star rating system. The safety assessment covers four areas: adult protection, child protection, pedestrian protection and safety assist [5]. Safety requirements needed to obtain a constant rating become stricter every year. While NCAP tests traditionally rely on tests of passive safety, active safety has become more and more important. Systems for accident avoidance and mitigation - such as AEB

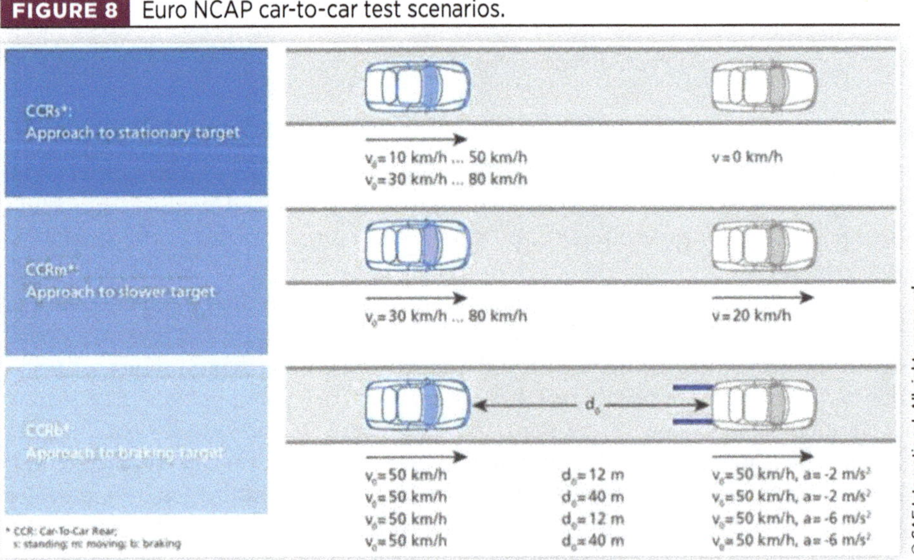

FIGURE 8 Euro NCAP car-to-car test scenarios.

(autonomous emergency braking) systems with autonomous braking and/or warning functions - have been included in the rating since 2014. AEB systems can be rated as:

1. Car-to-Car Rear Stationary (CCRs): the ego-vehicle or vehicle under test (VUT) is driving towards a stationary vehicle target (Euro NCAP vehicle target, EVT). This example is discussed in more detail below.
2. Car-to-Car Rear Moving (CCRm): the vehicle under test approaches a slow target vehicle.
3. Car-to-Car Rear Braking (CCRb): the vehicle under test follows a target vehicle that suddenly starts decelerating at a rate of either 2 m/s^2 or 6 m/s^2

The test suites of AutomationDesk concentrate on the Car-to-Car Rear test cases. AutomationDesk therefore provides a suitable test environment. These test suites make it possible to test and validate ECU functionalities according to the Euro NCAP test catalogs through simulation such as HIL simulation [12]. AutomationDesk performs the tests automatically, then records and compares the test results. The tool also provides ready-to-use tests for automatically parameterizing and executing the Euro NCAP test scenarios. The test results are evaluated according to the Euro NCAP criteria and documented in a report.

The test suites include the method of signal-based test description. The benefit of using this kind of test description is that the maneuver control for the NCAP tests is reproducible. This significantly increases the deterministic nature of the entire test procedure. The terms that are used for implementation and for evaluating the NCAP tests can be evaluated and carried out in real-time. In addition, the tolerance tube for all captured signals can be set to evaluate the performance of the system under test (SUT) in the Euro NCAP testing scenario (Figure 9). The evaluation can quickly be adapted

FIGURE 9 Signal-based test description and signal captures of NCAP test maneuvers.

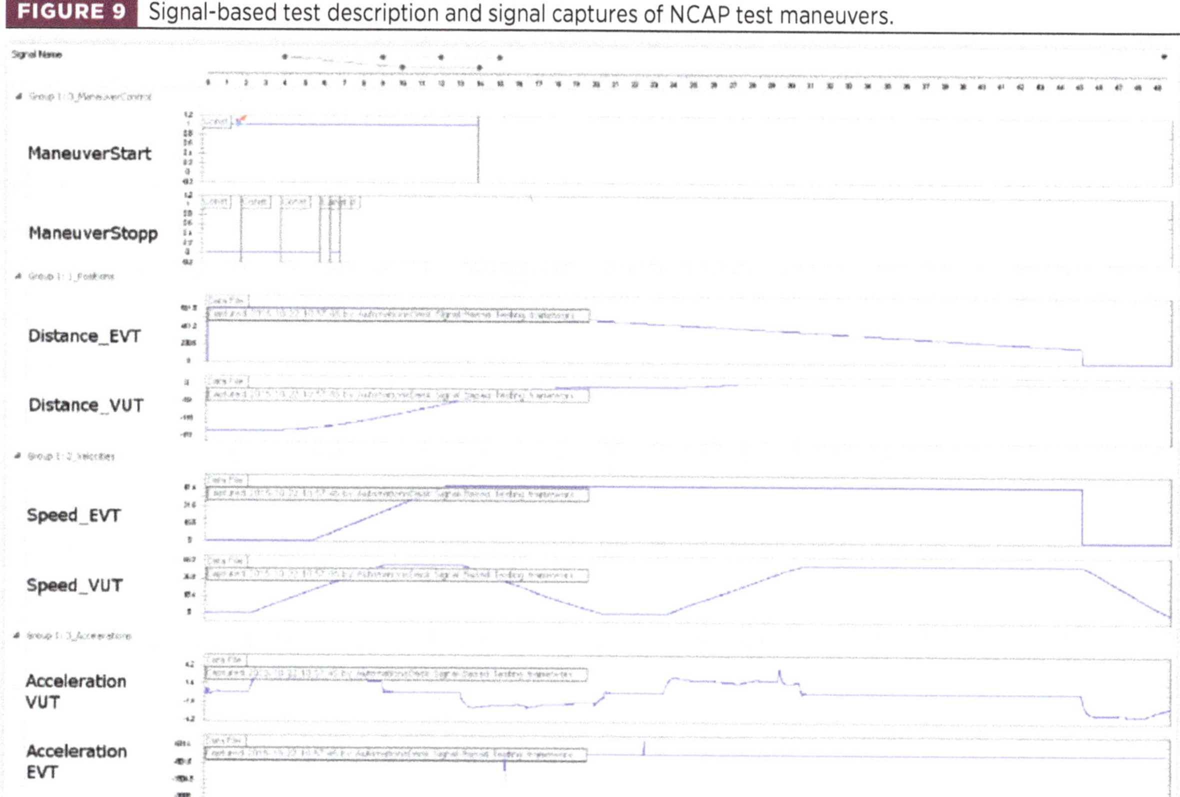

accordingly to different parameter sets for the SuT or the real-time model or different scenarios in which the SuT is tested. The evaluations are not only time based they depend on the situations of the real-time simulation e.g. the position or speed of the cars within the test environment.

This use case shows that it is possible to use a test automation tool that provides an intuitive test descriptions to automate test suites like the Euro NCAP test catalog without neglecting the strict real-time conditions for test execution and test evaluation. It is therefore possible to combine increased usability with a high test coverage.

Summary/Conclusions

The two aspects of pre-qualified COTS test automation tools mentioned in this paper and the test automation tool's level of usability can be factors that increase productivity in development processes by offering possibilities to lower the effort for and hurdles towards efficiently using a test automation tool in the verification process. That is, if the vendor of the test automation tool provides appropriate support.

A pre-qualified software tool that can be integrated into an existing verification process lowers the effort for classifying and qualifying this software tool in accordance to ISO 26262, because most of the necessary steps to integrate the tool in accordance to the standard have already been completed by the tool's vendor. By providing a certificate for the test automation tool and also a reference workflow that describes how to integrate the test automation tool into existing verification processes, tool vendors support users of test automation tools with substantial know-how and assistance.

Usability is a productivity increasing factor because it gives users the ability to work with the tool productively without expert knowledge of the tool. They can concentrate on their actual tasks. Furthermore, the start of using the test automation tool is not slowed down by additional effort for incorporating it into any other specific tool.

When all of these factors are combined, a significant productivity enhancement is possible. This enhancement is necessary to counteract the additional effort in verification processes, which is triggered by the automotive industry trend of gaining complexity and the increasing number of safety-related development projects that involve E/E systems.

Contact Information

Dr. Andreas Himmler
Senior Product Manager Hardware-in-the-Loop Testing Systems
dSPACE GmbH
Rathenaustraße 26
33102 Paderborn
Germany
ahimmler@dspace.de

Dr. Klaus Lamberg
Lead Product Manager Prototyping and Validation Software Tools
dSPACE GmbH
Rathenaustraße 26
33102 Paderborn Germany
KLamberg@dspace.de

Tino Schulze
Lead Product Manager Hardware-in-the-Loop Testing Systems
dSPACE GmbH
Rathenaustraße 26
33102 Paderborn
Germany
TSchulze@dspace.de

Jann-Eve Stavesand
Product Engineer Hardware-in-the-Loop Testing Systems
dSPACE GmbH
Rathenaustraße 26
33102 Paderborn
Germany
JStavesand@dspace.de

Definitions/Abbreviations

COTS - Commercial off-the-shelf
HIL - Hardware-in-the-loop
TÜV - Technischer Überwachungsverein
STCA - Software tool classification
TI - Tool impact
TD - Tool error detection level
TCL - Tool confidence level
E/E - Electric/electronic
ISO - International Organization for Standardization
IEC - International Electrotechnical Commission
ASAM - Association for Standardization of Automation and Measuring Systems
XIL - X-in-the-loop
AEB - Autonomous emergency Braking
Euro NCAP - European New Car Assessment Programme
SUT - System under test
API - Application programming interface
PC - Personal computer

References

1. dSPACE GmbH, "AutomationDesk—Safety Manual," 2015.
2. "Test Automation Software," 2015.
3. dSPACE GmbH, "TÜV SÜD Certifies AutomationDesk according to the ISO 26262 and IEC 61508 Standards," *dSPACE Magazin*, 2015.
4. Duisberg, A., "ISO 26262—A Reliable Standard from a Product Liability Viewpoint?," *4. Euroforum Jahrestagung. Stuttgart*, September 12-14, 2012.
5. Euro NCAP, October 23, 2015. http://www.euroncap.com/en.
6. "Gesetz über die Haftung für fehlerhafte Produkte" (Produkthaftungsgesetz - ProdHafG), 1989.

7. Himmler, A., Lamberg, K., and Beine, M., "Hardware-in-the-Loop Testing in the Context of ISO 26262," SAE Technical Paper 2012-01-0035, 2012, doi:10.4271/2012-01-0035.
8. IEC 61508, "Functional Safety of Electrical/Electronic/Programmable Electronic Safety-Related Systems," International Standard, International Electrotechnical Commission, 2010.
9. ISO 26262:2011, "Road Vehicles—Functional Safety," International Standard, International Organization for Standardization, 2011.
10. ISO 26262-8, ISO 26262-8, "International Standard Road Vehicles—Functional Safety—Part 8: Supporting Process," 2011.
11. Klindt, T., "Haftungsrechtliche Wirkung von technischen Normen," Euroforum-Konferenz ISO 26262, 2014.
12. Krügel, K., Sänger, N., Horyds, G., and Deutenberg, G., "Innovative Stragetgies for Validation and Testing in Function Development," *IAV Symposium*, 2014.
13. Lamberg, K., "Using dSPACE AutomationDesk for Safety-Related Applications," 2011.
14. Rasche, R., Brückner, C., and Neumerkel, D., "Ready, Set, Go! Measuring, Mapping and Managing with XIL API 2.0," *ASAM International Conference*, Dresden, Germany, December 3-4, 2013.

CHAPTER 3

Safe and Secure Development: Challenges and Opportunities

Jana Karina von Wedel and Paul Arndt
INVENSITY GmbH

The ever-increasing complexity and connectivity of driver assist functions pose challenges for both Functional Safety and Cyber Security. Several of these challenges arise not only due to the new functionalities themselves but due to numerous interdependencies between safety and security. Safety and security goals can conflict, safety mechanisms might be intentionally triggered by attackers to impact functionality negatively, or mechanisms can compete for limited resources like processing power or memory to name just some conflict potentials. But there is also the potential for synergies, both in the implementation as well as during the development. For example, both disciplines require mechanisms to check data integrity, are concerned with freedom from interference and require architecture based analyses. So far there is no consensus in the industry on how to best deal with these interdependencies in automotive development projects. SAE J3061 introduces a process framework for Cyber Security development that is intentionally very similar to that for Functional Safety as defined in ISO 26262. While these parallel frameworks help to identify interdependencies and show that aligned processes are possible, a joint process seems unreasonable due to the vastly different implementation frameworks and methods. Using concrete examples, we show problems that can arise if Functional

CITATION: von Wedel, J. and Arndt, P., "Safe and Secure Development: Challenges and Opportunities," SAE Technical Paper 2018-01-0020, 2018, doi:10.4271/2018-01-0020.

Safety and Cyber Security processes are not properly aligned and integrated into the overall development process. Based on this we then propose steps towards coordinated safety and security processes that can prevent such problems and show how such an approach at the same time allows to benefit from synergies.

Introduction

All safety-relevant systems are also security relevant. One of the main tasks of Cyber Security when dealing with such systems is to prevent attackers from being able to negatively influence the Functional Safety of a vehicle [1]. To reach its goals of authentication, authorization, availability, confidentiality, and integrity Cyber Security must also consider in which ways attackers might exploit safety mechanisms. A natural concern is a possibility of triggering safety degradations to reduce or disable functionalities, thereby negatively influencing the availability. Functional Safety, on the other hand, must ensure that Cyber Security mechanisms or faults in them cannot themselves have a safety-critical impact.

Understanding these interdependencies and implementing effective mechanisms to deal with them becomes more and more complex of a problem as the scope and connectivity of driver assist functions increase. The goal of higher and higher degrees of automation of driving functions and eventually fully autonomous driving necessitate a high degree of communication both between the different systems within a vehicle as well as between the vehicle and its environment [2]. At the same time, it increases the demands on Functional Safety. Both because functions must be allowed a higher degree of influence on safety-relevant systems within the vehicle to enable such functions, and because the driver cannot be counted on to be able to mitigate negative effects of the functions in the same manner and to the same degree as before [3].

Both the Functional Safety and the Cyber Security communities are aware of these challenges. The ISO 26262 was extended in the second edition to also cover the relationship between Functional Safety and Cyber Security. On Cyber Security side, the process framework introduced in SAE J3061 was intentionally aligned with that of ISO 26262 to ease the introduction of Cyber Security processes where those for Functional Safety are already in place [1].

While the process frameworks in these two standards are the same, a closer examination shows that the actual activities necessary to implement them differ greatly. The mentioned interdependencies between the two disciplines on the one hand and the vast differences regarding methods, tools, and mechanisms that they use on the other hand must both be considered when deciding how to handle the aspects of Functional Safety and Cyber Security during development.

The focus of this paper is first to propose an integrated approach to dealing with them in a manner that helps to prevent conflicts while at the same time enabling synergies. To this end, we first compare ISO 26262 and SAE J3061 in more detail in Chapter 2 to show the interrelations and dependencies between Functional Safety and Cyber Security for automotive systems. In Chapter 3 we use concrete examples to show how these interrelations can cause conflicts and problems during development. In Chapter 4 we propose steps towards aligned processes that aim at identifying potential conflicts early on and therefore enable their prevention. We will further show how such aligned processes not only help to prevent conflicts but at the same time allow for benefiting from synergies. The paper concludes with a summary in Chapter 5.

ISO 26262 and SAE J3061: Interrelations and Dependencies

On a high level, the goal of Functional Safety and Cyber Security is the same: Reduce risk to an acceptable level. Both disciplines agree that elimination of risk is not possible [1, 4]. However, it must be ensured that steps are taken and mechanisms are used that help to reduce risk and are feasible and reasonable concerning the related development effort, costs, and functional impact, and that the remaining risk is in line with the state of the art.

Both ISO 26262 and SAE J3061 prescribe a top-down, design-driven approach. Functional Safety and Cyber Security respectively are to be considered throughout the whole development project and build into the design rather than added onto a finished system at the end of the development [1, 4]. The process framework used in both standards, therefore, covers the whole product lifecycle from concept phase to decommissioning. Such an all-encompassing approach is only feasible if the parties involved at the various levels of a development project (OEM, Tier 1, Tier 2, etc.) work together. Both standards intend that their application is possible at different development levels and include guidance for how to define the interfaces between those levels concerning Functional Safety and Cyber Security respectively.

At first glance, the scopes of safety and security activities seem strictly separated: Safety mechanisms protect against faults from within the system, security mechanisms protect against attacks from outside of the system. Even though Functional Safety and Cyber Security target different aspects of system integrity, they do have many interrelations. The probably most obvious one being that whenever a system is safety-relevant, Cyber Security must analyze whether it might be possible for an attacker to negatively influence the Functional Safety of the system and implement mechanisms to prevent this if necessary.

Acknowledging this fact that all safety-relevant systems are also security relevant, both standards explicitly consider the relationship to the respectively another discipline. In ISO 26262 this is done via a separate Annex that lists potential interactions between the two disciplines. SAE J3061 not only lists preventing attackers from compromising the safety of an automotive system as one of the central goals of Cyber Security but also uses the process framework introduced in ISO 26262 as a basis for defining process steps towards the achievement of secure systems. This was explicitly done to allow organizations with safety processes in place to better align their security processes with these safety processes [1].

Consequently, the main steps for achieving Functional Safety according to ISO 26262 and Cyber Security according to SAE J3061 are parallel, as also depicted in Figure 1. Both approaches start by defining the system boundaries and interfaces to determine the scope of considerations (Item Definition/Feature Definition), then based on this determine the associated dangers and evaluate their risk (HARA/TARA) and derive goals for the development based on this analysis (Safety Goals/Cyber Security Goals). In the next steps, first, a functional concept for reaching these goals that can include both process and functional measures (Functional Safety Concept/Functional Cyber Security Concept) and subsequently a corresponding technical concept refining the functional measures to concrete technical solutions (Technical Safety Concept/Technical Cyber Security Concept) are created.

FIGURE 1 Some of the major work products requested by ISO 26262/their conceptual counterpart in SAE J 3061 within the V model for development.

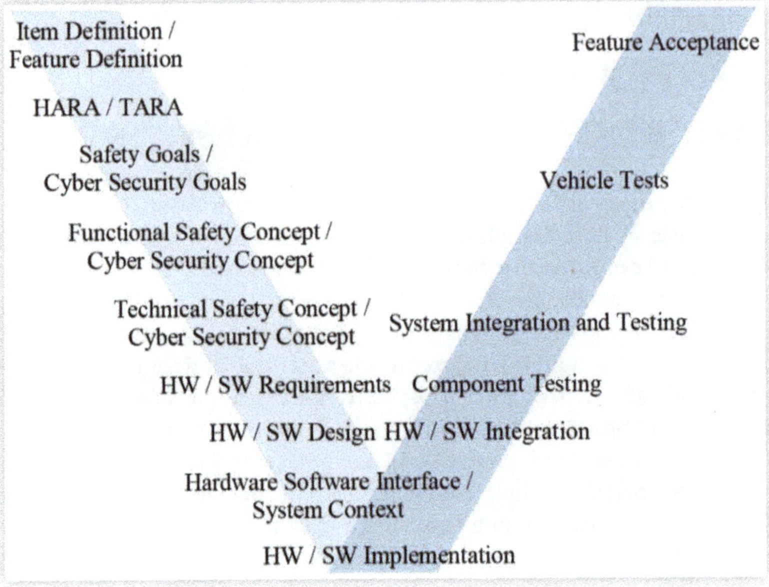

Item Definition / Feature Definition
HARA / TARA
Safety Goals / Cyber Security Goals
Functional Safety Concept / Cyber Security Concept
Technical Safety Concept / Cyber Security Concept
HW / SW Requirements
HW / SW Design
Hardware Software Interface / System Context
HW / SW Implementation
HW / SW Integration
Component Testing
System Integration and Testing
Vehicle Tests
Feature Acceptance

The similarities continue throughout the sub-system development. From the technical concepts, requirements for hardware and software are derived. The interaction between hardware and software components and any additional requirements that might derive from it are considered as well (Hardware Software Interface/System Context).

In-depth system and design analyses are performed throughout the development process both on system and subsystem level to identify any potential additional relevant faults or vulnerabilities and to judge the sufficiency of the implemented measures. The verification and validation activities further include requirements based tests and integration tests both on system and sub-system level, stress tests, and tests aimed at checking the system response in case of faults or attacks (Fault Injection Testing/Penetration Testing).

All these activities are planned at the beginning of the development project, and responsibilities for the steps are assigned (Safety Plan/Cyber Security Program Plan). For performing these activities, Functional Safety and Cyber Security also rely on the same supporting processes being in place, including configuration management, documentation management, and change management with their respective work products.

Regular checks are performed to ensure that the processes defined in the planning were adhered to, and the resulting work products have the required quality (Safety Audit, Verification Reviews and Confirmation measures/Gate Reviews). The development activities are concluded by creating a structured argument supported by evidences for how the risk was reduced to an acceptable level (Safety Case/Cyber Security Case).

Functional Safety and Cyber Security activities continue throughout production, the whole lifetime of the vehicle in the field including service and maintenance up to the decommissioning. This includes not only checking whether requirements and instructions for these steps that were defined during development are adhered to, but also dealing with safety or security relevant incidents in the field (Instructions Regarding Field Observations / Incident Response Process).

Even though at this high level the methods and goals of Functional Safety and Cyber Security respectively are comparable, the actual implementations of the work products differ vastly. The differences in focus and conditions lead to very different challenges being faced by the two disciplines, therefore necessitating different approaches during development. In the following, this will be demonstrated using the examples of Hazard Analysis and Risk Assessment (HARA) and Threat Analysis and Risk Assessment (TARA).

This example was chosen since HARA and TARA are two of the first major work products requested by ISO 26262 and SAE J3061 respectively and people working in either the field of Functional Safety or that of Cyber Security therefore tend to be familiar with the respective analysis process. It should however be explicitly noted that the differences in the challenges faced by the two disciplines continue throughout the whole development process and product lifecycle and vary for the individual work products created during the different stages.

HARA and TARA

During these analyses, the risk associated with hazards respectively threats relevant to the item or system at hand is to be evaluated. While these goals are very similar, they dictate fundamentally different scopes for the analyses. On Functional Safety side, this scope is relatively easy to define. Only E/E faults of the item at hand itself as well as foreseeable misuse by the user must be considered while other systems and items may be assumed to work correctly. Therefore, a high confidence that all relevant hazards are covered can be achieved by analyzing the possible faults of the item and their potential impact with related hazards.

Cyber Security, on the other hand, must consider attacks both by the user and third parties. Not only attacks using interfaces of the system at hand with the environment, but also the use of other systems as stepping stones for attacks on the system and vice versa must be taken into account. This means that relevant threats depend not only on the system itself but also on the functionality of connected systems. These differences in scope are illustrated in Figure 2.

Since capabilities of potential attackers are not always known or may change over time, it is much harder to gain confidence that all relevant threats were considered than it is for Functional Safety concerning hazards. This can be compensated to some degree by the fact that while their importance and concrete impact of course vary depending

FIGURE 2 Differences in analysis scope: Functional Safety considers risks originating in faults within the vehicle system while Cyber Security looks at risks caused by attacks from outside of the system.

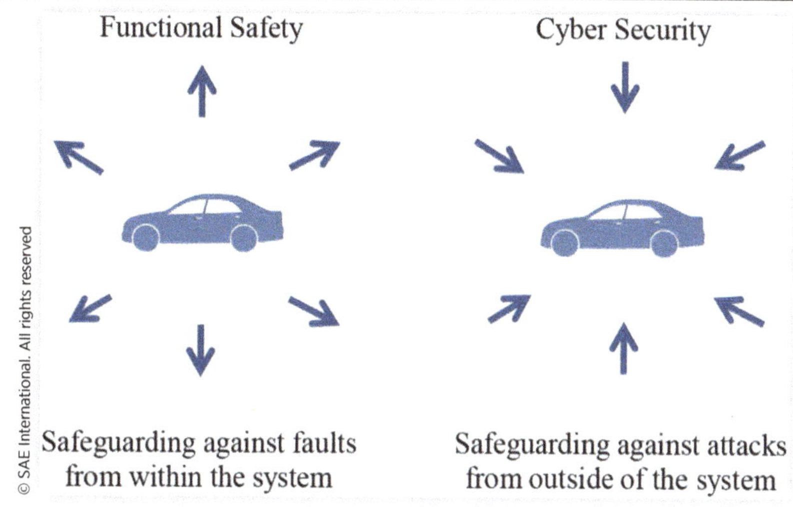

CHAPTER 3

on the system at hand, the relevant types of threats and security mechanisms tend to be the same for all kinds of systems. The same cannot be said for Functional Safety where the relevant hazards and safety mechanisms vary greatly between different items since they depend to a much greater degree of the functionality that the item provides.

The fact that the capabilities of potential attackers change over time also causes vast differences in the validity of HARA and TARA. As long as the item does not change, the relevant hazards and their ASIL ratings are extremely unlikely to change either. While there might be minor differences in the individual ratings for exposure, severity and controllability, e.g. due to the intended market of the vehicle and related laws and regulations, or due to assumptions made regarding the group of users when judging controllability, a given hazard of a given item is likely to receive the same ASIL rating today as it will 5 years from now.

On Cyber Security side, however, the relevant threats to be considered as well as their evaluation in the TARA can change much quicker and much more significantly. When judging the relevance of a potential threat and evaluating the risk associated with it, it must among others be taken into account which relevant knowledge is currently how easily available and how likely potential attackers will have access to specialized equipment that might be used during an attack. Consequently, the risk associated with a threat can increase very quickly as potential attackers can gain new abilities or equipment, previously unknown or not well-known methods can become public knowledge in the relevant communities, etc. At the same time, risks might also decrease over time. If for example a threat related to extracting confidential information currently has a high impact due to pending patents, the impact and therefore overall risk will significantly decrease once the patent has been granted. A given threat for a given system might hence be evaluated as much more, or much less, critical just a few months from now than it is today. This makes it necessary to update a TARA whenever new information becomes available.

Not only can the risk evaluation change significantly over time, but these many aspects that must be considered generally make it hard to determine the probability of an attack related to a given Cyber Security threat. So far, no accepted methods are available for their quantification. Determining the probability factor for the risk associated with a hazard during a HARA is comparatively much easier. It depends only on the operational situation for which the hazard is being evaluated. Numerous statistics are available to help with this step and examples for the probability of operational situations are even given directly in ISO 26262.

The issue of a lack of accepted scales and methods to be used during a TARA also applies to the impact factor. On Functional Safety side, the severity of injuries that the user or other traffic participants might suffer as an effect of the hazard, as well as the likelihood that they will be able to control the situation to avoid such harm are evaluated. Widely available statistics on traffic accidents as well as own trials regarding controllability can be used to judge these aspects. The existing mapping between the rating of injuries that might occur during traffic accidents on the abbreviated injury scale (AIS) and the severity ratings proposed by ISO 26262 further eases this process. Cyber Security, on the other hand, must evaluate not only the impact of threats on safety but also that on several other factors like availability, privacy, and critical infrastructure. Quantifying these aspects can prove difficult. Not only immediate monetary losses but also long-term effects on costs and revenue should be taken into account, for which relevant statistics to base a rating on are unlikely to be available in sufficient number.

Despite those numerous differences between HARA and TARA they are also closely related. When conducting the HARA, faults, and reactions of security mechanisms must be considered as potential causes of hazards. This also includes attacks on the system that can be seen as foreseeable misuse and their handling by the planned security mechanisms. During the TARA, on the other hand, it must be examined whether planned safety mechanisms and the defined safe states could be exploited by an attacker, e.g., to influence the availability of assist functions negatively. Naturally, the possible impact of threats and associated attacks on functional safety must be checked as well. It should be noted here that there is no direct correspondence between the ASIL rating of a safety goal and the risk associated with a threat that can lead to a violation of this safety goal. This is because only the impact factor of the risk is influenced by the ASIL rating while the probability factor is independent of it.

Conflict Potentials

The differences in scopes, approaches, and methods used by Functional Safety and Cyber Security respectively as well as the interrelations and dependencies between the two disciplines that were outlined above can lead to conflicts not only during development but also concerning the implemented mechanisms. Preventing those conflicts from actually occurring is not always straightforward. Conflicting safety and security goals or objectives, or limited resources that both disciplines must utilize can make it challenging to find effective and efficient solutions for implementing systems that are both safe and secure.

Conflicting Goals

An example that shows how safety and security goals can conflict is that of the central door locking functionality. In case of a crash, the doors of the car should be unlocked to allow injured passengers to exit the vehicle and first responders to be able to get to them quickly. If there is uncertainty as to whether a crash occurred or not, the safe reaction is therefore to unlock the doors. The central door locking function also serves a security purpose though, namely to keep unauthorized people out of the car. From a security point of view, the best reaction in case of doubt regarding the current state of the vehicle is therefore to have the doors locked.

An attacker could exploit the safety mechanism that unlocks the doors in case a crash is assumed to have occurred. If no valid status information from the airbag is received for a certain period, the safety mechanism will usually trigger an unlocking of the doors. If an attacker can influence the communication bus over which the airbag status information is sent, he or she can either manipulate the messages from the airbag control unit or attempt to overload the bus such that the messages do not arrive in time. In the case of manipulation of the message containing the status information, it is not even necessary to manipulate the message such that it indicates a crash. Instead, it is sufficient to change it such that the receiver of the message can detect that the data was corrupted, e.g., because the attached checksum / CRC does not fit the message content. An overloading of the communication bus to delay the arrival of the status messages and thereby triggering an unlock tends to be less complicated for the attacker since it does not require the manipulation of the messages. Such attacks have been shown to be possible in numerous vehicles, frequently using ECUs with remote connections like Bluetooth or Wi-Fi handling convenience functions as stepping stones [5].

Developers must be aware of such conflicts and interdependencies in order to be able to define strategies for dealing with them. In the example of central door locking the risk might be reduced by modifying the safety mechanism such that a missing or corrupted airbag signal is no longer a sufficient precondition for unlocking the doors. Signals for which certain values can be expected in case of a crash could be used for defining additional conditions. For example, a condition on the progression of the vehicle speed which is usually available to the control unit implementing the central door locking anyways could be used.

When designing mechanisms that aim to prevent exploitation of safety mechanisms by attackers, it should naturally be ensured that they do not prevent any safety mechanisms from working as intended. In some cases it might not be possible to completely prevent conflicts between safety and security concepts though, making it necessary to explicitly prioritize either safety or security. Since preventing negative effects on the Functional Safety of a system is one of the goals of Cyber Security, prioritizing safety seems to be the more natural choice. However, also the risk associated with the affected hazard(s) and threat(s) should be taken into account. If the risk related to a Cyber Security threat is much higher than that of Functional Safety hazard it might also be decided to prioritize security in some conflict cases. This shows that in order to decrease the overall risk related to both safety and security to an acceptable level, each discipline must be aware of the goals and strategies of the respectively other one such that joint strategies can be developed.

Conflicts in Mechanisms

Even when the goals of Functional Safety and Cyber Security are consistent, conflicts can arise when defining the safety and security mechanisms for a function. These conflicts can be on a principle level. Safety-relevant functions tend to rely on signals from other ECUs in the vehicle not only for their functionality but also for their safety mechanisms. They are used for plausibility checks and to create redundancy. From a safety perspective, when additional information that can be used for such a purpose is available from a different ECU than the ones that the mechanism is already using signals from, using the information from that ECU is a preferable solution. This is due to the fact using signals from different ECUs in a calculation reduces the chance of common cause faults influencing several of the input signals.

From a Cyber Security perspective, increasing the number of ECUs in the vehicle from which information is received means opening up new potential avenues for attacks. Both the manipulation of the system at hand via the connected ECUs as well as the use of the own system as a stepping stone for attacks on the connected ECUs must be considered. These considerations are not limited to attacks directly targeting the safety integrity of a system, but must also include attacks that can have unintended side effects that might influence functional safety as well as such that could compromise other Cyber Security goals. This also means that an update of the TARA might be necessary if the decision is made to receive additional messages from potentially additional ECUs for the use in safety mechanisms.

When inputs for safety functions from other systems within the vehicle are missing, delayed, or detectably corrupted, a safety reaction or degradation is frequently triggered. This means that attacks that while not violating Safety Goals might violate Cyber Security Goals are possible by preventing messages from being sent or delaying them significantly, or by corrupting them in any way that can be detected by the receiver. The first might be possible by flooding the communication bus with additional messages. Since many safety-relevant functions within the vehicle react to missing or unreliable information by triggering safety reactions, such an attack could have numerous and sometimes hard to predict effects that are relevant to Cyber Security.

The extent of such effects naturally varies from vehicle to vehicle depending among others on the provided driver assist functions, the network topology, and the way in which safety-relevant functions are implemented. One example was already mentioned in the previous sub-chapter, the possibility of the doors being unlocked by the central door locking functionality in case the airbag status information is missing or corrupted. Another is that of missing or corrupted information about the vehicle speed potentially leading to degradations or functions being disabled concerning numerous systems and functions like steering system, exterior lights, braking system, cruise control, charging system of electric vehicles, or parking assist.

This does of course not mean that increasing the number of interfaces and connections within a vehicle should generally be avoided. In fact, similarly to Functional Safety, Cyber Security can frequently profit from the availability of additional information that can be used in order to detect an attack. It does however once again show that safety and security concepts must be properly aligned to prevent negative side effects for the respective other discipline.

Fault Tolerance Time

Not only conflicting goals, objectives, or strategies can lead to challenges in developing safe and secure systems, but also limited resources like processing power, memory, or simply time, that safety and security mechanisms might compete for. One such example is that of safety-critical systems that have a short fault tolerance time associated with certain hazards. The fault tolerance time is defined as the time between the occurrence of a fault that would if not mitigated, lead to a hazard and the moment at which the system enters a safe state, thereby preventing any further safety-relevant effects of the fault [4].

In the following, we will use the example of an electronic power steering (EPS) system. One of the hazards commonly considered for this type of system is that of an unintended actuation of the steering system. Depending on the power of the motor of the steering system, the dynamics of the vehicle, and the current driving situation such an unintended actuation can have a significant effect on the intended path of the vehicle within a very brief time. Consequently, the fault tolerance time associated with this hazard is very short, usually between 12 and 20 ms.

When a fault that could lead to an unintended actuation is detected the safe reaction is to reduce the maximal torque of the motor, potentially down to zero which corresponds to a switching-off of the steering support. In the case the current support torque generated by the motor is high, the hardware reaction alone can use up a significant portion of this fault tolerant time. Additionally, it must be considered that it might not be possible to detect the fault causing the unintended actuation immediately, but that in the worst case this could also take several milliseconds.

It is conceivable that an attacker might try to manipulate safety-relevant inputs received by the EPS to intentionally trigger a safety reaction, in this case, a reduction of the steering support. Such an attack would then violate the Cyber Security goals of ensuring availability and functional integrity. Therefore, security mechanisms should be devised to prevent this kind of attack.

However, measures like authentication and encryption take time in processing which must be considered when distributing the fault tolerance time for safety hazards among the involved components and functions. As outlined above, the fault detection and mitigation could in the worst case already completely use up the time budget if it is small. Any additional delay due to security checks, even if those add only one or two milliseconds, could, therefore, lead to the system exceeding the fault tolerance time and thereby potentially becoming uncontrollable for the driver.

Additional to the necessity for coordination of safety and security mechanisms, this example also shows that both should be included in the design of the overall system from the beginning of the development. Especially when the reaction time of hardware is a factor, requirements towards the overall system and the hardware design could arise additionally to software requirements. For example, using hardware incorporating cryptographic accelerators like a Hardware Security Module (HSM) or the use of Secure Hardware Extensions (SHE) might be necessary to overcome such issues. If such requirements are not yet known in the concept and early design phase of an automotive development project, their implementation is likely to lead to significant additional costs.

Adjusting for Evolving Security Threats

Cyber Security risks evolve as attacker's motivations and capabilities change. The Cyber Security Concept of an automotive system can, therefore, involve defenses against techniques that may not be (fully) understood at the time the system is created. Frequent updates of the security mechanisms are therefore necessary [1]. As discussed above, changes in the implemented security mechanisms may have an impact on Functional Safety since both the mechanisms themselves as well as faults in them must be considered as potential sources of hazards.

In safety engineering, a common approach to safety mechanisms is never to change them if it can at all be prevented. The longer a safety mechanism remains unchanged and the more cars it is used in, the higher the confidence in its correct function and capability to sufficiently reduce the risk of harm to life or limb.

Therein lies another potential conflict between Functional Safety and Cyber Security. When changes in security mechanisms are necessary the question arises as to whether changes should be restricted such that they do not have a significant impact on functional safety or whether also changes with significant impact should be allowed, even though this means losing the advantage of increased confidence in safety mechanisms thanks to the available field data and test depth.

Steps towards an Integrated Approach

The outlined interrelations and dependencies between Functional Safety and Cyber Security and the resulting conflict potentials highlight the need for strategies to deal with the two disciplines in an integrated manner during automotive development projects. As shown via examples, even though the used process framework is the same, the actual activities and methods necessary to implement Functional Safety and Cyber Security respectively are very different. Straight-out combining the two disciplines and placing responsibility for them in the same position therefore seems likely to not be the most efficient and effective solution.

Instead, we believe that, as long as they are not both handled as an integral part of the overall development process, Functional Safety and Cyber Security should be treated as separate disciplines during development while maintaining close coordination. Since safety and security mechanisms do not only influence each other but also the overall system design not only coordination between the two disciplines but also their integration into the overall development processes is essential for an efficient development. Communication paths need to be established early in the development project to ensure coordination of the technical concepts. Regular meetings between the project members responsible for the two disciplines as well as between them and other stakeholders in the system, software, or hardware architect can help to identify potential conflicts early on and to find solutions that can be beneficial to both disciplines.

The examples of HARA and TARA as well as Safety and Cyber Security Concept that were discussed above show that both disciplines need results from the respective other as input to be able to ensure that all relevant hazards respective threats are sufficiently covered by the technical concept. Since this dependency holds in both directions, existing processes and methods might have to be adapted to cover the needs of both disciplines. Instead of performing a HARA or TARA as a one-time effort, an iterative process might have to be adapted such that the results of the HARA can be considered in TARA and vice versa. Appropriate checkpoints need to be built into the product lifecycle at which not only the status of work products of the individual disciplines is checked, but also the coordination across disciplines.

The set-up of security processes in organizations that already have established safety processes might further be helped by some lessons-learned from Functional Safety, e.g., already consulting an expert during the acquisition phase to prevent unrealistic expectations, having a single communication channel towards the customer or supplier on safety/security topics, etc.

This particularly also concerns the cooperation between OEMs and suppliers. Experience in the development of safety relevant automotive systems has shown that in order to ensure the Functional Safety of a vehicle both parties must work together. While them OEM has a unique view of the overall vehicle architecture and the interdependencies between systems within the vehicle, suppliers have a much more detailed know-how regarding the systems they are developing and can therefore e.g. better judge the impact of external influences or come up with more efficient mechanisms meeting the needs of the overall vehicle concept. For this same reason, cooperation is also needed to ensure the Cyber Security of a vehicle. Detailed knowledge both regarding the overall vehicle concept as well as the design of single systems might be necessary in order to judge whether additional measures are necessary to protect against a given threat. Due to the fact that contrary to hazards for Functional Safety, Cyber Security threats can evolve over time, this need for cooperation might be even greater regarding security topics than it is for safety issues.

Established processes for cooperation across development levels regarding Functional Safety can be used as a basis for establishing similar processes for Cyber Security. For example, a document similar to the Development Interface Agreement (DIA) as requested by ISO 26262 in which responsibilities and information exchange related to Functional Safety are agreed between two development parties could be established for Cyber Security.

Both Functional Safety and Cyber Security will face new challenges due to increased networking both within vehicles and with the infrastructure as well as more highly automated or autonomous driving functions, making the need for increasing synergies and reducing conflicts even greater. To this end, it is advisable to find vocabulary, development tools, mechanisms, architectural building blocks, design patterns (see also [6]), etc. that work for both domains. In the following, we will, therefore, discuss some areas in which synergies are possible, both during the development as well as in the resulting implementation.

Data Integrity

One of the areas in which the targets of Functional Safety and Cyber Security align is that of data integrity. The integrity of data both in transit as well as in storage must be protected, or if such a protection fails, the loss of integrity must be detected to prevent undesirable and potentially critical safety system and vehicle reactions. For Functional Safety, the focus is on protection against systematic and random errors caused by malfunctions or unintended interference. Methods commonly used to achieve this include message counters and checksums or CRC for messages, double storage of safety-relevant data, ECC, and the use of MPUs. Cyber Security aims to additionally protect data against targeted, intended, and possibly malicious manipulation. Frequently key based Message Authentication Codes (MAC) are used to achieve this goal.

In many current automotive systems, the respective safety and security mechanisms work in parallel. This is despite the fact that numerous mechanisms cannot distinguish between accidental and intentional manipulation of the data. Therefore, security mechanisms also cover cases that are checked by safety mechanisms and vice versa. In the following, this is highlighted using the examples of MAC and CRC.

When using MAC, the sender uses a cryptographic key to generate a code that is attached to a sent message. The same key is available at the receiver who also uses it to generate a code and then compares it to the one that was attached to the message by the sender. This way any corruption of data in transit, either accidental or intentional, can be detected while at the same time verifying the identity of the sender. CRC, which is most commonly used for ensuring data integrity in transit in safety mechanisms, uses a similar approach. Both sender and receiver perform a polynomial division on the message and the results of these computations are compared.

There are however two fundamental differences between the two approaches. One is that the key for generating MACs is designed such that an attacker cannot derive the key from messages he or she intercepts while the polynomial used for CRC could well be determined if enough messages are intercepted. The other concerns the probability with which different types of errors are detected. Using MACs, any corruption of data can be detected with the same probability, i.e. no error classes are distinguished. With CRC on the other hand, errors concerning a small number of bits can usually be detected with a higher probability than those affecting many bits. Frequently it is even possible to not only detect but also correct single bit errors based on CRC. This is due to the fact that the polynomial used for CRC is frequently chosen based on the assumption that single bit errors are binomially distributed, meaning the higher the number of bits affected the lower the probability of an error is assumed to be.

While the fact that the used polynomial could be derived if enough messages are intercepted makes CRC inadequate from a security perspective, MAC could be used for safety purposes if the assumption of equal probability for all kinds of errors is acceptable (see also [7]). By analyzing the needs concerning data integrity by safety and security mechanisms side by side, one can choose mechanisms that best suit the needs of both. This enables a reduction of implementation and testing effort because fewer mechanisms are used. It can also reduce the need for additional information and qualifiers that need to be added to bus messages and thereby reducing their length and the busload.

Freedom from Interference

Another aspect that both Functional Safety and Cyber Security are concerned with is that of freedom from interference. Safety aims to separate safety-relevant software components which realize ASIL requirements from those that realize only QM requirements as much as possible to prevent systematic or random interference from those non-safety relevant components. If the computational infrastructure allows for it, this could, for example, be realized via virtualization.

Such a separation does not only help to achieve the safety goals of a system but can be beneficial for security as well. Security mechanisms designed to protect against threats that can negatively influence Functional Safety can be devised in a more targeted and focused manner when proper separation is given. Cyber Security itself is further concerned with freedom from interference between components which need to communicate with the environment and those that do not have to. The mechanisms used for safety mechanisms are suitable for this security purpose as well. Using the same kind of mechanisms can reduce the related overall implementation and testing effort.

Security Mechanisms as Safety Basis

Several common security mechanisms can help to reduce the need for additional safety mechanisms or at the very least improve the confidence in the sufficiency of the implemented safety mechanisms. One example for this is secure flashing or secure boot. These mechanisms ensure that only authenticated software is flashed and loaded. They are introduced by Cyber Security to prevent the introduction of malware using which further attacks on the system at hand might be possible. Also from a safety perspective, it is necessary to know which software is run on a system to be able to ensure that the necessary safety mechanisms are included and active. The Safety Concept can make use of implemented security mechanisms, e.g., if those prevent certain configurations from occurring it is no longer necessary to devise strategies for how to safely react if unexpected configurations are used.

Another example, which while maybe not being able to reduce the need for safety mechanisms is still able to increase the confidence in their sufficiency, is that of the authentication of hardware components. Security mechanisms ensure that only original parts are used, thereby preventing negative publicity or lowered customer satisfaction due to a potentially lower performance with non-original parts as well as financial losses. At the same time, a negative impact on the Functional Safety of the system is prevented. If non-original parts were used in the system that was not considered during the safety analyses performed during development, the occurrence of safety hazards due to random hardware faults might be higher than deemed acceptable.

Verification and Validation

Even though many of the same methods are used for the verification of both safety and security mechanisms, e.g., equivalence class tests, boundary value tests, stress tests, and static code analyses, the different focuses, and objectives of the V&V activities of the two disciplines prevent a joint approach. Code that is valid or correct from a safety perspective may still have Cyber Security vulnerabilities and vice versa.

Especially when it comes to validation testing and identifying potential additional or remaining risks, some fundamental differences can be seen. Functional Safety uses fault injection testing and vehicle tests to validate that the implemented safety mechanisms are sufficient to reduce the risk of harm to an acceptable level by preventing the occurrence or mitigating the effects of hazards. The respective activities on Cyber Security side not only aim to validate that the implemented security mechanisms mitigate the previously identified vulnerabilities sufficiently, but additionally try to identify additional, previously unknown vulnerabilities, e.g., by breaking, bypassing, or tampering with the implemented security mechanisms. The scope of the penetration testing performed for this purpose is much less defined as that for fault injection testing on the safety side.

When it comes to analyses, however, synergies are possible. Safety and security analyses have the same basis, namely architecture and requirements. Further, both types of analyses focus on propagation throughout the system, of faults in the one case and manipulations in the other. Although the focus of the analyses is different, the steps taken during their performance overlap. This makes joint safety and security analyses possible.

Both software safety analysis and software vulnerability analysis are performed on the software architecture and focus on data and control flow to identify the critical paths regarding safety respectively security. It is then checked whether the mechanisms planned or implemented so far are sufficient to reduce the risk associated with these critical paths to an acceptable level. While the evaluation of criticality and sufficiency will differ between safety and security analysis, the underlying data and control flow analysis are the same, and it is, therefore, sufficient to perform it only once. Further, since the influence of attacks on the Functional Safety of the system must be considered by Cyber Security, the information gathered during the evaluation phase of the software safety analysis is a valuable input for the respective phase of the software vulnerability analysis.

Similarly, a deductive safety analysis like FTA and a vulnerability analysis, e.g., using an attack tree, are both based on an analysis of the system architecture. Again, it is possible to lay the groundwork for both analyses in a joint step and to further use the results of the safety analysis as additional input for the security analysis. It has even been proposed to completely combine the two analyses in one methodology [8].

Summary/Conclusions

Based on a comparison of the processes for achieving Functional Safety according to ISO 26262 and Cyber Security according to SAE J3061 as well as of the activities necessary to implement the respective process steps, we have highlighted the many interrelations and dependencies that exist between the two disciplines. Using concrete examples of automotive systems, we have shown how they can lead to conflicts both during development as well as regarding the implementation of the respective system at hand. Given that these conflict potentials are likely to increase rather than decrease as increased networking both within vehicles and with the infrastructure as well as more highly automated or autonomous driving functions pose challenges for both disciplines, we concluded that an integrated development approach for Functional Safety and Cyber Security is needed.

Therefore, we proposed steps towards such an integrated approach that focuses on a close coordination of the two disciplines both with each other and with the overall development throughout the project to detect potential conflicts early on as well as on finding methods, tools, and strategies that can benefit both. We then discussed some of the areas in which such synergies are possible and how they can be benefited from.

Contact Information

{jana.wedel, paul.arndt}@invensity.com

Definitions/Abbreviations

ASIL - Automotive Safety Integrity Level
CRC - Cyclic Redundancy Check

ECC - Error Correction Code
e.g. - Exempli gratia
ECU - Electrical Control Unit
EPS - Electronic Power Steering
HARA - Hazard Analysis and Risk Assessment
HSM - Hardware Security Module
i.e. - Id est (in other words)
ISO - International Standardization Organization
MAC - Message Authentication Code
MPU - Memory Protection Unit
QM - Quality Management (lowest rating for the safety criticality of functions in ISO 26262)
SAE - Society of Automotive Engineers
SHE - Secure Hardware Extension
TARA - Threat Analysis and Risk Assessment
V&V - Verification and Validation
Wi-Fi - IEEE 802.11x

References

1. SAE International Surface Vehicle Recommended Practice, "Cybersecurity Guidebook for Cyber-Physical Vehicle Systems," SAE Standard J3061, Rev. January 2016.
2. Organization for Economic Co-Operation and Development (OECD), "Automated and Autonomous Driving," International Transport Forum Policy Papers, April 1, 2015, doi:10.1787/5jlwvzdfk640-en.
3. Horwick, M. and Siedersberger, K., "Strategy and Architecture of a Safety Concept for Fully Automatic and Autonomous Driving Assistance Systems," *2010 IEEE Intelligent Vehicles Symposium*, June 2010, doi:10.1109/ivs.2010.5548115.
4. ISO (International Organization for Standardization), "Road Vehicles—Functional Safety," DIS/ISO 26262, Rev. 2016.
5. Koscher, K., Czeskis, A., Roesner, F., Patel, S. et al., "Experimental Security Analysis of a Modern Automobile," *IEEE Symposium on Security and Privacy*, 2010, doi:10.1109/sp.2010.34.
6. Amorim, T., Martin, H., Ma, Z., and Schmittner, C., "Systematic Pattern Approach for Safety and Security Co-Engineering in the Automotive Domain," *Computer Safety, Reliability, and Security* (2017): 329-342, doi:10.1007/978-3-319-66266-4_22.
7. Glas, B., Gebauer, C., Hänger, J., Heyl, A. et al., "Automotive Safety and Security Integration Challenges," *Automotive-Safety & Security* (2014): 13-28.
8. Mamdouh, E., "Vulnerability Tree Analysis Versus Fault Tree Analysis—Combined Security\Safety Analysis Approach," Presented at *EUROFORUM ISO 26262 2017*, USA, October 9-11, 2017.

CHAPTER 4

Steering Control Based on the Yaw Rate and Projected Steering Wheel Angle in Evasion Maneuvers

Yifan Ye, Jian Zhao, Jian Wu, Bing Zhu, and Yang Zhao
Jinlin University, ASCL

Weiwen Deng
Beihang University

When automobiles are at the threat of collisions, steering usually needs shorter longitudinal distance than braking for collision avoidance, especially under the condition of high speed or low adhesion. Thus, more collision accidents can be avoided in the same situation. The steering assistance is in need since the operation is hard for drivers. And considering the dynamic characteristics of vehicles in those maneuvers, the real-time and the accuracy of the assisted algorithms is essential.

In view of the above problems, this paper first takes lateral acceleration of the vehicle as the constraint, aiming at the collision avoidance situation of the straight lane and the stable driving inside the curve, and trajectory of the collision avoidance is derived by a quintic polynomial. Based on the control of the steering wheel angle by the optimal preview control algorithm, the differential braking control is carried out by using the feedbacks of yaw rate and the projected steering wheel angle information to improve the accuracy of trajectory tracking and the stability of the ego vehicle in evasion maneuver.

Simulation analysis based on the vehicle dynamic software (ASM) is conducted in typical maneuvers. And the results show that the coordinated steering algorithm can further improve vehicle tracking

CITATION: Ye, Y., Zhao, J., Wu, J., Zhu, B. et al., "Steering Control Based on the Yaw Rate and Projected Steering Wheel Angle in Evasion Maneuvers," SAE Technical Paper 2018-01-0030, 2018, doi:10.4271/2018-01-0030.

accuracy and vehicles' stability when using the same collision avoidance trajectory under the limit of designed lateral acceleration. It can partly decrease the influence of error of steering systems since the use of projected steering wheel angle and contribute to the convergence of lateral accelerations.

Introduction

There are approximately 2 million U.S. police-reported rear-end crashes in 2014 according to the analysis of Insurance Institute for Highway Safety [1]. Therefore, diver assistant systems in collision avoidance get more and more concentration from researchers and authorities, with the improved demands of vehicle safety. In general, the ego vehicle can avoid the collision by braking, steering or the combination of them when there is an imminent collision, so the assistant functions focus on active braking and steering.

Autonomous Emergency Braking (AEB) systems is a driver assistant system which conducts automatic braking to avoid or mitigate the imminent collisions when drivers are inattention or distracted. With maturity of technology, AEB has been brought into the market, such as City Safety of Volvo, Toyota Safety Sense and so on. It has achieved good results in reducing the collision accidents. Related researches find that vehicles with low-speed AEB were involved in 38 percent fewer police-reported rear-end striking injury crashes than similar vehicles without AEB in a meta-analysis of benefits in six mainly European countries [2].

According to researches of human drivers' behaviors, the steering activities increase with shorter Time-to-collision (TTC), and shorter TTC often means that the situation is emergency and dangerous [3]. However, it is difficult for drivers to control vehicles properly in those maneuvers. And the improper steering operations will still lead to collision accidents, even cause instability of the vehicle. Therefore, driver assistant function is in demand in those maneuvers, where AEB is not applicable. In addition, respondents think the evasion steering support is significant and more important than common AEB in low speed on the market based on a user survey [4]. Accordingly, the researches of the steering support in those maneuvers have gained great momentum in recent years, such as the Emergency Steer Assist (ESA) of Continental [3], the Evasive Steering Support (ESS) of Bosch [5], and the Evasive Manoeuvre Assistance System (EMAS) of Volvo [6].

There are two modes in steering assistance in evasion maneuvers, which are steering support and autonomous steering. Steering support systems detect the traffic environment through the environment sensors. Once there is a potential collision risk and the driver chooses to steer to avoid the collision, the systems will support the driver to control the vehicle move along an optimal trajectory. Autonomous steering is often triggered when the driver is distracted without any steering or braking input, the collision cannot be avoided by fully braking, and there is a feasible path to evade the collisions by steering. There are three main problems in the control, environment sensing, trajectory planning and trajectory tracking, this paper focuses on the latter two.

Researchers have proposed many trajectory planning methods, such as interpolating curve [7, 8], the potential field [9], and rapidly-exploring random tree [10]. In evasion maneuvers, real-time of the algorithm, continuity of the curve and the curvature is crucial for tracking precision and the vehicles' stability. Therefore, the interpolating curves have gotten many applications, since their low computational cost and continuity. Trajectory tracking of autonomous vehicles can be divided into three stages. In the early stage, the pure tracking control for robot based on kinematic model and positional error has been applied [11]. Since dynamic characteristics of automobile have significant influence on the tracking accuracy, the control based on simplified vehicle dynamic model and the preview algorithm are adopted, and get widely applications [12]. Feedback regulation based on states of the vehicle such as position and heading angle are adopted to further enhance the stability [13]. The optimal algorithms such as MPC have been proposed in recent years, since the advantage of its comprehensive consideration of vehicle kinematic and dynamics constrains [14].

In this paper, considering the demand of real-time and the influence of the dynamic characteristics in evasion maneuver. Quantic polynomial is adopted in the trajectory planning. The general expressions of the trajectory are derived in the two typical maneuvers, which are that vehicles travel in straight lane and circle lane. Then, the optimal preview algorithm is used for calculated the desired steering wheel angle. Furthermore, considering the error of simplified dynamic model and limitation of the tire friction circle when there are larger lateral accelerations, feedbacks of yaw rate angle based on the projected steering wheel angle are added into the controller and achieved by differential braking.

This paper is organized as follows: the scheme frames of steering assistance in evasion maneuvers are described and analyses in section 2. In section 3, trajectory planning are developed when the ego vehicle at imminent collisions in straight lane or circle lane. Then, the schemes of trajectory design and trajectory tracking are developed. Simulations in typical maneuvers are conducted, and results are presented in section 4. Finally, conclusions are summarized in section 5.

System Description

The Steering control in evasion maneuver developed in this paper is a driver assistance function which can help the driver avoid the imminent collisions by steering. The control mode is autonomous steering, which is the latter one depicted above. There are three initial condition, an imminent collision, a feasible path under the limit of maximum lateral acceleration, and the driver is districted without any input of steering or braking. The steering controller will control the vehicle steer to avoid the collision combining with the differential braking. For the consideration of safety, the diver has the highest priority. When the conflicts between the driver's input and the controller input reach a certain threshold, the system will exit.

The architecture of steering control in this paper is shown in Figure 1. The system can be divided into three functional modules. Environment Sensing, Trajectory Planning and Trajectory Tracking. Environment Sensing is responsible for detecting the traffic information, and assessing the possibility of collisions combining with the motion state of the ego vehicle. When the result is there is a forthcoming collision, Trajectory Planning module will try to calculate the feasible path to avoid the collisions according to the traffic information from Environment Sensing. If there is a reasonable path under the limitation of lateral accelerations, Trajectory Planning module will generate a reference path. Then, the Trajectory Tracking module will control the vehicle to track the reference path and avoid the collisions. This paper focuses on the Trajectory Planning and Trajectory Tracking.

FIGURE 1 Architecture of coordinated steering controller in evasion maneuver.

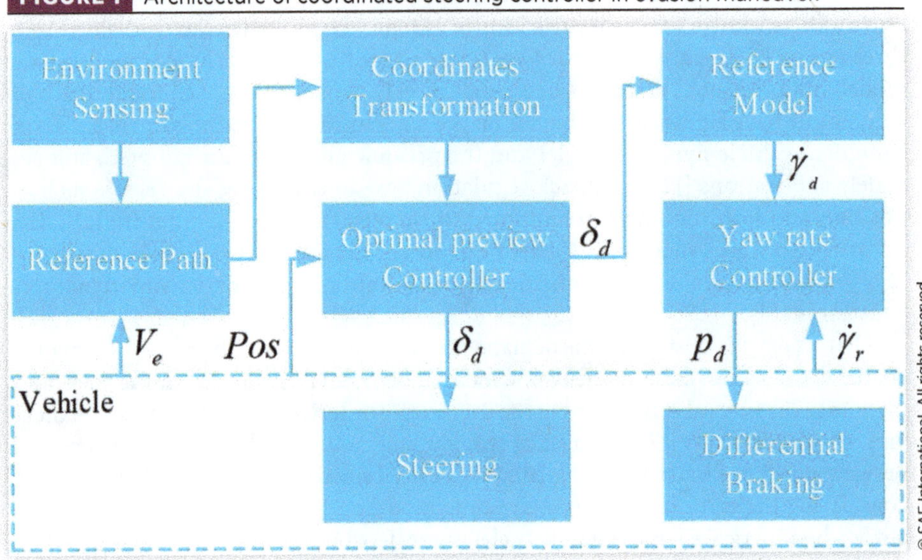

To satisfy the requirements of collision avoidance, this paper adopts the quantic polynomial to calculate the reference path. Since it has the continuous and bounded curvature, which can avoid an abrupt steering wheel angle input and limit the lateral accelerations. In this paper, we chose two typical traffic condition, which are the ego vehicle travels in straight lane or travels in a circle to derive the generic function of the reference path with the constraints at the start point and the end point. As a result, the relationship between lateral acceleration and longitudinal distance is established. So, the reference path can be designed according to lateral acceleration and velocity of the ego vehicle.

For the trajectory tracking control in evasion maneuvers, this paper adopts the optimal preview algorithm as a fundamental algorithm. This algorithm simplifies vehicles to the 2 degree of freedom model. However it have some error, especially when the vehicle travels with a large lateral acceleration, which just happens in evasion maneuver. Reference to working principle of Electronic Stability Control (ESC), which also take same 2 degree of freedom (DOF) simplified model for the reference model and make the respond of the vehicle more in line with the driver's input, this paper takes the projected steering wheel angle as input, and try to eliminate the error between the reference model and the ego vehicle by differential braking. In this way, the control errors caused by the inaccuracy of the reference model will decrease. As a result, the trajectory tracking will be more accurate and the possibility of collision avoidance will increase.

Control Algorithm Development

Trajectory Planning of Collision Avoidance

The feasible trajectory is fundamental condition for the success of collision avoidance. In general, the planning of trajectory need to considerate vehicle stability and occupant comfort. This article chooses two typical scenarios, the vehicle travels in a straight lane and in a circle road. According to the characteristics of boundary constrains under corresponding scenarios. The trajectory of collision avoidance is derived in Cartesian coordinate system and polar coordinate system, which is represented by quintic polynomials. After the general expressions are obtained, the trajectory design scheme is introduced.

STRAIGHT ROAD

When the ego vehicle travels in a straight road and there is a potential collision between the ego vehicle and leading vehicles on the same lane, the ego vehicle can achieve certain lateral displacement by steering to avoid the occurrence of the collision, as shown in Figure 2.

Assuming that the ego vehicle travels at a constant speed along the straight lane before avoidance and after it. The following boundary conditions can be summered to satisfy constrains of the position, slope and curvature in these evasion maneuvers.

$$x_0 = 0,\ y_0 = 0,\ \dot{y}_0 = 0,\ k_0 = 0 \tag{1}$$

$$x_e = x_e,\ y_e = y_e,\ \dot{y}_e = 0,\ k_e = 0 \tag{2}$$

The spline has closed expressions, and it is easy to ensure the continuity and boundedness of curvature [15]. According to the boundary conditions, the quintic polynomial is minimum orders spline to satisfy them. Therefore, the trajectory curve equation is assumed to be as follows.

$$y(x) = a_0 + a_1 x + a_2 x^2 + a_3 x^3 + a_4 x^4 + a_5 x^5 \tag{3}$$

Since the curvature of y(x) is

$$k = \frac{\ddot{y}}{\left(1+\dot{y}^2\right)^{3/2}} \tag{4}$$

Therefore, $k = 0$ is equal to $\ddot{y} = 0$.

Finally, the curve satisfies boundary condition of (1) (2) can be derived as followed Expression (5).

$$y(x) = y_e\left[10\left(\frac{x}{x_e}\right)^3 - 15\left(\frac{x}{x_e}\right)^4 + 6\left(\frac{x}{x_e}\right)^5\right] \tag{5}$$

CIRCLE ROAD

When the ego vehicle is on the corners and there is an imminent collision. Similar to the situation when it travels in a straight line, ego vehicle can achieve a radial displacement to avoid the collision as shown in Figure 2. In this situation, this paper assumes

FIGURE 2 Collision avoidance trajectory in straight road.

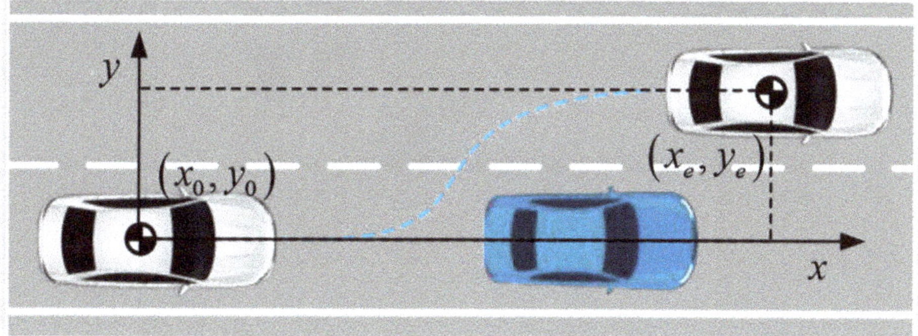

that the ego vehicle and lead vehicle travel on concentric circles. In order to make the expression more concise, the trajectory is derived in the polar coordinate system (ρ, θ), and the origin of it is in the center of the circle road.

Assuming that the ego vehicle is in a steady circling in the curve before the collision and after it. Then, the following boundary conditions can be summered to satisfy constrains of the position, slope and curvature in these evasion maneuvers.

$$\theta_0 = 0,\ \rho_0 = R_0,\ \dot{\rho}_0 = 0,\ k_0 = 1/R_0 \tag{6}$$

$$\theta_e = \beta,\ \rho_e = R_e,\ \dot{\rho}_e = 0,\ k_e = 1/R_e \tag{7}$$

Therefore, the general form of the trajectory which is defined by the polar radial ρ and the polar angle θ can be expressed as

$$\rho(\theta) = b_0 + b_1\theta + b_2\theta^2 + b_3\theta^3 + b_4\theta^4 + b_5\theta^5 \tag{8}$$

The curvature of $\rho(\theta)$ is

$$k = \frac{\rho^2 + 2\dot{\rho} - \rho\ddot{\rho}}{\left(\rho^2 + \dot{\rho}^2\right)^{3/2}} \tag{9}$$

Thus, $\ddot{\rho} = 0$ can be derived from $k = 1/\rho$ and $\dot{\rho} = 0$.

The desired curve satisfies the constraints of (5) (6) can be derived as follow.

$$\rho(\theta) = R_0 + R_d\left[10\left(\frac{\theta}{\beta}\right)^3 - 15\left(\frac{\theta}{\beta}\right)^4 + 6\left(\frac{\theta}{\beta}\right)^5\right] \tag{10}$$

Where R_d is the radical displacement.

$$R_d = R_e - R_0 \tag{11}$$

FIGURE 3 Collision avoidance trajectory in circle road.

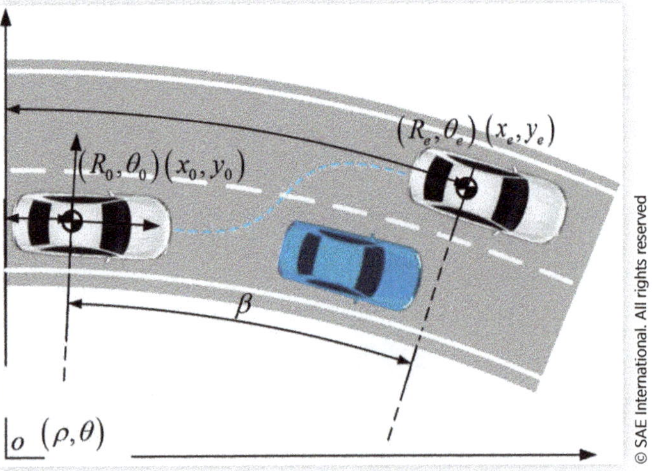

To get a uniform expression, the desired curve can be transformed to Cartesian coordinate system on the ego vehicle as follow.

$$\begin{cases} x = \left(\rho(\theta) - R_0\right) \cdot \cos\theta \\ y = \rho(\theta) \cdot \sin\theta \end{cases} \tag{12}$$

TRAJECTORY DESIGN SCHEME

According to the derivation above, the relationship between the path maximum curvature and the end point is established. In this paper, the end point is set on the left lane of the ego vehicle in order to reduce the disturbance to traffic when avoid collision. Therefore, the lateral displacement is set as a standard lane width, 3.75 m. In this way, the path only changes with longitudinal position of the end point as shown in Figure 4. It can be found that the curvature is continuous and has a boundary. Considering the lateral acceleration can be written as Function (13), where v denotes the velocity in tangent direction of the avoidance trajectory, which is assumed to be constant in this paper. Therefore, this study set a maximum lateral acceleration as a limitation to get the

FIGURE 4 Path and curvature with different longitudinal displacements.

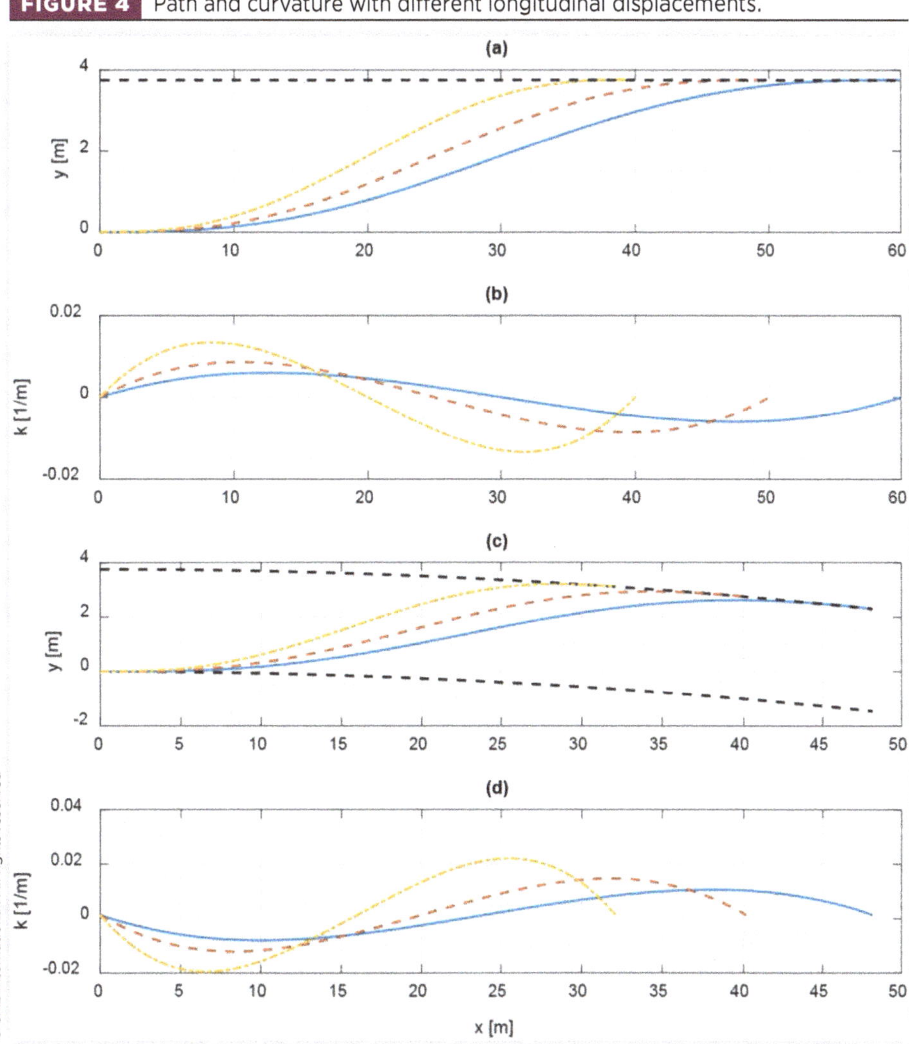

desired path. When the desired path does not conflict with the path of the targets and there is an imminent collision, the steering controller will steer the ego vehicle to avoid the collision.

$$a_y = v^2 \cdot k \tag{13}$$

Trajectory Tracking

REFERENCE MODEL

In this paper, 2 DOF vehicle model with the freedoms of yaw angle and lateral displacement is used as the reference dynamic model. The schematic diagram of it is shown in Figure 5.

Select the state variable x as the following expression

$$x = \left[y, v_y, \dot{\gamma}, \gamma \right] \tag{14}$$

Where y is the lateral displacement, v_y is the velocity in the Y-axis, $\dot{\gamma}$ is the yaw rate and γ is the yaw angle. Then, taking steering wheel angle as the input u, the model can be written in state space form as Equation (15).

$$\dot{x} = Ax + Bu \tag{15}$$

Where,

$$A = \begin{bmatrix} 0 & 1 & 0 & v_x \\ 0 & -\dfrac{2C_F + 2C_R}{mv_x} & -\dfrac{2C_F l_a - 2C_R l_b}{mV_x} - v_x & 0 \\ 0 & -\dfrac{2C_F l_a - 2C_R l_b}{I_z v_x} & -\dfrac{2C_F l_a^2 + 2C_R l_b^2}{I_z v_x} & 0 \\ 0 & 0 & 1 & 0 \end{bmatrix} \tag{16}$$

$$B = \begin{bmatrix} 0 \; 0 \; \dfrac{2C_F}{mi_m} \; \dfrac{aC_F}{I_z i_m} \end{bmatrix}^T \tag{17}$$

Where, C_F and C_R are lateral stiffness of the front tires and the rear tires. v_x is velocity in the X-axis, m is the total mass of vehicle, I_z is the moment of inertia Z-axis, and i_m is the mean of steering ratio.

FIGURE 5 Diagram of a vehicle 2 DOF dynamics model.

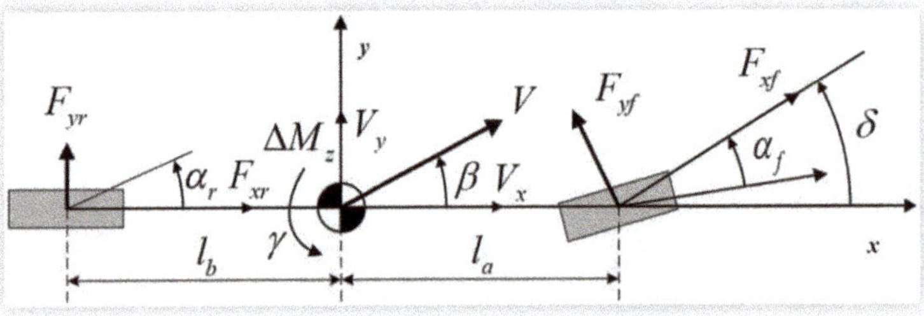

STEERING WHEEL CONTROL

The steering wheel control is based on the optimal preview control, which is proposed by MacAdam [12]. That algorithm adopts the quadratic sum of lateral deviation between the vehicle position and the desired trajectory in certain preview time as an optimization goal, and the steering wheel angle is calculated to minimize the deviation under a constant value. As it is shown in Figure 6.

According to the reference dynamic model above, when we take the lateral position of the vehicle as the output, the output equation can be written as (18).

$$y = Cx + Du \tag{18}$$

Where,

$$C = \begin{bmatrix} 1 & 0 & 0 & 0 \end{bmatrix},\ D = 0 \tag{19}$$

Then, the output y can be write as the function of time.

$$y(t) = Ce^{(At)}x_0 + C\int_0^t e^{(A\eta)}Bud\eta \tag{20}$$

Where x_0 is the longitudinal position of vehicle when $t = 0$. Assuming the desired lateral position is $y_{ref}(t)$, the objective function is

$$J = \frac{1}{T_0}\int_0^{T_0}\left[y_{ref}(t) - y(t)\right]^2 W(t)dt \tag{21}$$

Where $W(t)$ is the weighting function, T_0 is the preview time. Furthermore, in discrete time $kT = T_0/m$. J can be described as

$$J = \sum_{k=0}^{m}\left(y_{ref}(kT) - Ce^{(AkT)}x_0 - C\int_0^{kT} e^{(A\eta)}Bud\eta\right)^2 W_k \tag{22}$$

To calculate the derivative of the target function with respect to the input variable u and make it to be zero, the optimal input of the steering wheel angle $u(0)$ can be derived. That is

$$u(0) = \frac{\sum_{k=1}^{m}\left(y_{ref}(kT) - Ce^{(AkT)}x_0\right)\left(C\int_0^{kT} e^{(A\eta)}Bd\eta W_k\right)}{\sum_{k=1}^{m}\left(\left(C\int_0^{kT}\exp(A\eta)Bd\eta\right)^2 W_k\right)} \tag{23}$$

FIGURE 6 Diagram of optimal preview model.

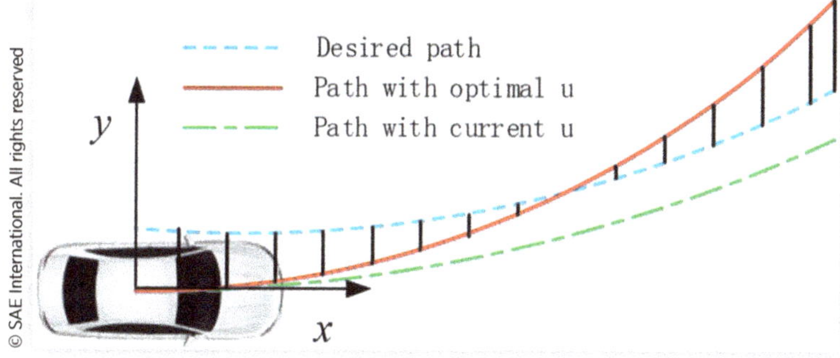

DIFFERENTIAL BRAKING CONTROL

In this paper, differential braking is adopted as a compensation measures for the incorrect or insufficiently steering input and the inaccuracy of the reference model when ego vehicles avoid collisions with a large lateral acceleration. The differential braking is based on the yaw rate information and the projected steering angle input.

As described above, the desired steering wheel angle input can be obtained from the desired trajectory and optimal preview controller. Therefore, the desired steering input δ_d can be directly used as the input of the reference model to calculate the reference yaw rate $\dot{\gamma}_r$, then the error between the reference yaw rate and the nominal yaw rate $\dot{\gamma}_n$ can be gotten, and a PID controller is designed to calculate the desired yaw moment M_{des} to eliminate the error.

$$\left(\frac{F_{xi}}{C_{xi}}\right)^2+\left(\frac{F_{yi}}{C_{yi}}\right)^2\le\left(\mu F_{zi}\right)^2 \tag{24}$$

Refer to the control principle of traditional ESC, the differential braking control chooses one wheel to brake at each time. According to the dynamic characteristic of tire, the maximum longitudinal force and lateral force are coupled as shown in Equation (24). Therefore, the lateral force may decrease when the longitudinal force increase with braking. The force diagram of vehicles when steering is shown in Figure 7. Hence, when the ego vehicle rounds a curve and the absolute value of nominal yaw rate $\dot{\gamma}_n$ is larger than reference yaw rate $\dot{\gamma}_r$, braking the front wheel on the outside is most effective. Since the moment produced by the increasing of longitudinal and the deceasing of lateral force is in the same direction. In the same way, braking the rear wheel on the inside is more effective when the absolute value nominal yaw rate $\dot{\gamma}_n$ is smaller than reference yaw rate $\dot{\gamma}_r$ [16]. And the details of chosen braked wheel according to the steering wheel angle and the error between $\dot{\gamma}_n$ and $\dot{\gamma}_r$ are shown in Table 1, $-\dot{\gamma}_t$ is the threshold of control.

FIGURE 7 Force diagram of vehicle when steering.

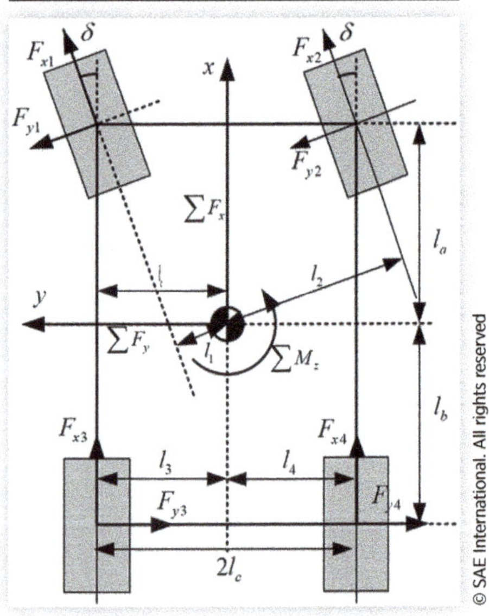

The desired wheel pressure p_d can be derived from the desired yaw moment and the position of braked wheel according to the Figure 7. Then, p_d can be achieved by controlling the hydraulic control unit of ESC. First, the desired longitudinal force F_y of the braked wheel is

$$F_y=\frac{M_{des}}{l_i}=\frac{k_p\left(\dot{\gamma}_r-\dot{\gamma}_n\right)}{l_i} \tag{25}$$

$$\begin{cases} l_1=l_c\cos\delta-l_a\sin\delta \\ l_2=l_c\cos\delta+l_a\sin\delta \\ l_3=l_4=l_c \end{cases} \tag{26}$$

Where l_c is half of wheelbase and l_a is distance between the front axis and the vehicle centroid. kp is the proportion coefficient of the controller. Then, the desired wheel pressure can be obtained.

$$p_d=\frac{2F_yR_w}{R_bA_bK_e}=\frac{2M_{des}R_w}{l_iR_bA_bK_e} \tag{27}$$

Where R_w denotes radius of the braked wheel, R_b is the effective radius of brake, A_b is the cross-sectional area of wheel cylinders, and K_e is the braking effective factor.

TABLE 1 Selection of the braked wheel

Yaw rate angle	Error $(\dot{\gamma}_r-\dot{\gamma}_n)$	Braked wheel
> = 0	$>\dot{\gamma}_t$	Rear Left
> = 0	$<-\dot{\gamma}_t$	Front Right
< 0	$>\dot{\gamma}_t$	Front Left
< 0	$<-\dot{\gamma}_t$	Rear Right

Simulation

In order to evaluate the algorithm described above, this paper established the control model in Simulink and conducted simulations in typical maneuvers based on the high-fidelity vehicle dynamic software Automotive Simulation Model.

The first simulation maneuver is that the ego vehicle travels at 120 km/h in a straight line and try to avoid the collision by steering under the limit of the maximum lateral acceleration, which is 6 m/s^2. The results of the simulation is shown in Figure 8. From the change of ego vehicle's position, it can be found that the algorithm above can track the desired path more accurately than the single optimal preview algorithm, which is represented by Pos Uncontrol. After further analysis, it can be found that the maximum lateral error between the reference path and the path without feedback of yaw rate is 0.47 m, and it is 0.33 m when applying the differential braking. If the boundary of the target vehicle is 2 m, the longitudinal distance to avoid the collision is 34.83 m. When choose to brake to avoid the collisions, the deceleration will be 15.9 m/s^2, which is impossible. Therefore, steering to avoid the collision needs a shorter distance than braking in

FIGURE 8 Compare of position, lateral accelerations and yaw rate in line.

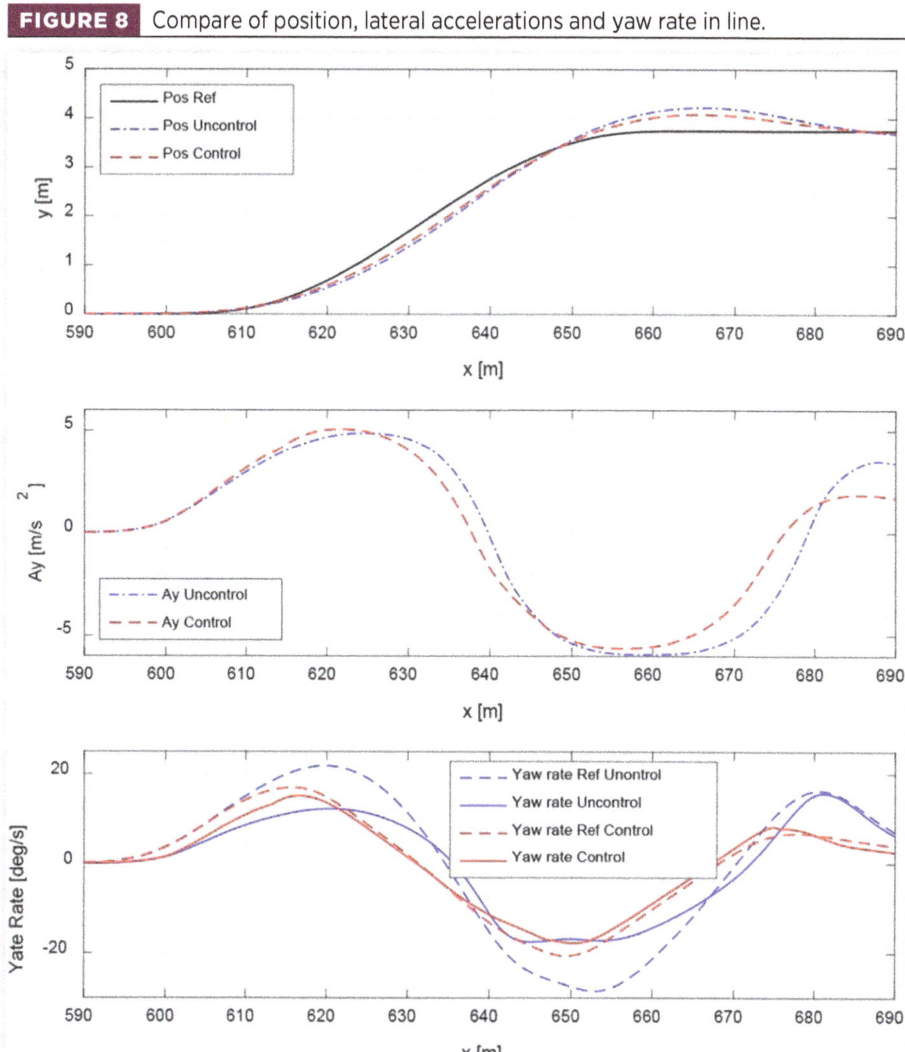

FIGURE 9 Compare of position and lateral accelerations in circle.

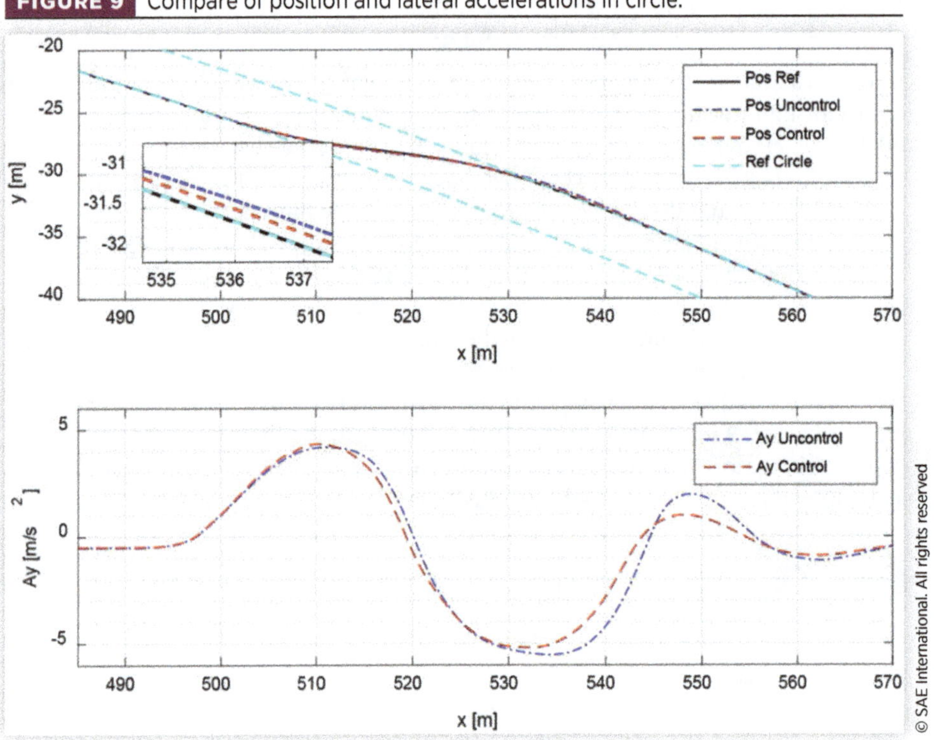

this maneuver. The actual lateral acceleration is limited in 6 m/s^2, which is in line with the intention of trajectory planning. In addition, it can be found that acceleration will convergent faster when the differential braking is applied, which is conducive to vehicle stability and comfort. The figure of the yaw rate shows that differential braking can follow the reference of 2 DOF model more accurately, and the maximum value is lowed since the accuracy of path tracking is increased and the maximum steering input is lower, which contributes to the stability in evasion maneuver.

Figure 9 is the simulation results when the vehicle travels at 72 km/h in circle and tries to avoid the collision by steering, the maximum lateral acceleration is also set as 6 m/s^2. The similar results can be gained. The maximum lateral error between the reference path and actual path is 0.27 m without differential braking, while it is 0.16 m when applying the differential braking. The lateral acceleration is continuous without abrupt change and can be limited according to the intension of trajectory planning when applying differential braking. The coordinated control scheme makes a faster convergence of lateral acceleration, which can enhance vehicle stability and the comfort of occupants in evasion maneuver.

In addition, simulations are conducted when the vehicle travels at 120 km/h, the limit of maximum lateral acceleration is 6 m/s^2, and the responsive steering wheel angle is 10% less than the projected steering angle. The Control1 in Figure 10 is the reference model input of differential braking is the responsive steering wheel angle, and Control2 is the projected steering wheel angle. As shown in the figure of position, we can find that differential braking with projected steering wheel angle get a better tracking performance. To be specific, the maximum error of lateral displacement is enhanced from 0.42 m to 0.36 m, and it is 0.59 m when without differential braking. And the lateral acceleration also gains a faster convergence. Table 2 is a summary of the maximum lateral error in each scenario.

FIGURE 10 Compare of position and lateral acceleration differential braking applied with projected steering wheel angle or not.

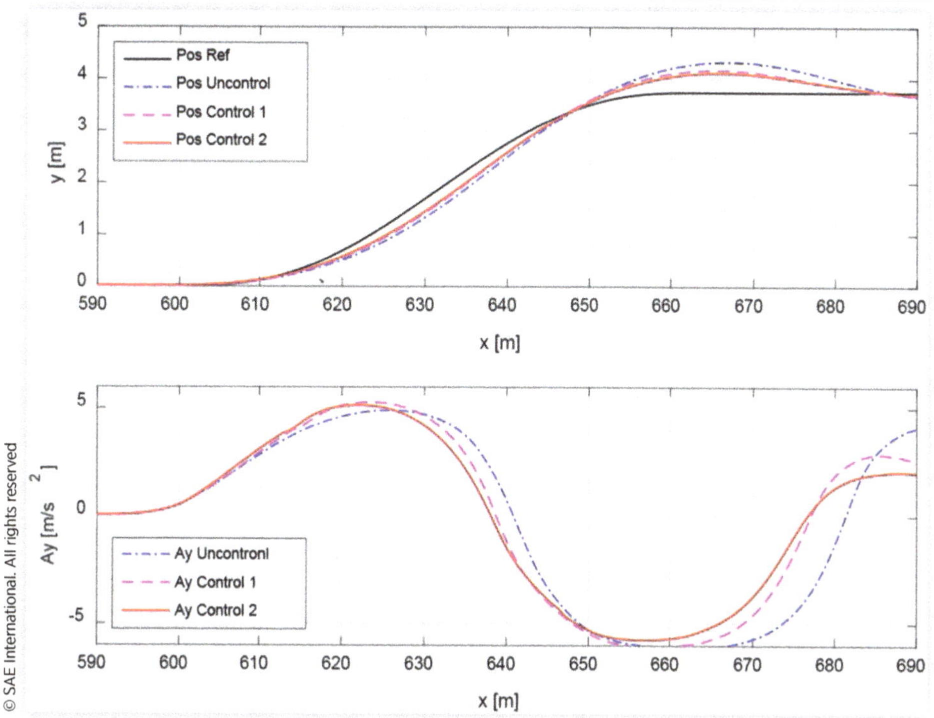

TABLE 2 Maximum error of lateral position in each scenario

Control mode	Line	Circle	90% steering
Optima steering wheel angle	0.47 m	0.27 m	0.59 m
Coordinated steering with projected steering wheel angle	0.33 m	0.16 m	0.36 m
Coordinated steering with responsive steering wheel angle	-	-	0.42 m

Conclusion

This study focuses on the steering control in evasion maneuvers. In order to satisfy the real-time requirements of trajectory planning, this paper adopts quintic polynomial to calculate the avoidance path. The curve has continuous curvature, which is conducive to trajectory tracking, and its boundedness makes it can be used as a limit of lateral acceleration. In Trajectory tracking, this paper combines the optimal preview algorithm and differential braking to get a better tracking performance. The results of simulation show that the coordinated algorithm can obtain a better tracking performance and stability of the ego vehicle. Taking the projected steering wheel angle as an input of the reference model in differential braking can compensate the error of steering system to a certain extent.

In the following study, the selection of threshold maximum lateral acceleration will be discussed according to road friction and the characteristic of drivers. General expressions of the avoidance trajectories in more maneuvers and the disturbance of steering avoidances on the traffic need to be further researched.

Contact Information

For questions or to contact the authors, please email to Professor **Jian Wu** at wujian@jlu.edu.cn.

Acknowledgments

This work is partially supported by National Natural Science Foundation of China (51575225, 51775235), and Jilin Province Science and Technology Development Plan Projects (20170101138JC, 20180201056GX).

References

1. Cicchino, J.B., "Effectiveness of Forward Collision Warning and Autonomous Emergency Braking Systems in Reducing Front-to-Rear Crash Rates," *Accident Analysis and Prevention* 99 (2017): 142-152.
2. Fildes, B., Keall, M., Bos, N., Lie, A. et al., "Effectiveness of Low Speed Autonomous Emergency Braking in Real-World Rear-End Crashes," *Accident Analysis and Prevention* 81 (2015): 24-29.
3. Eckert, A., Hartmann, B., Sevenich, M., and Rieth, P.E., "Emergency Steer & Brake Assist: A Systematic Approach for System Integration of Two Complementary Driver Assistance Systems," *Proceedings of the 22nd International Technical Conference on the Enhanced Safety of Vehicles*, Washington, DC, 2011.
4. Robert Bosch GmbH, *Chassis Systems Control Driver Assistance Systems—How Much Support do German Drivers Want?* (Heilbronn: Robert Bosch GmbH, 2012).
5. Fausten, M., "Accident Avoidance by Evasive Manoevres," *Proceedings of the 4th Tagung Sicherheit. Durch Fahrerassistenz TVSD Munich*, April 15-16, 2010.
6. Gurov, A., Sengupta, A., Jonasson, M., and Drugge, L., "Collision Avoidance Driver Assistance System Using Combined Active Braking and Steering,"*AVEC'14, 12th symposium on Advanced Vehicle Control*, Tokyo, Japan, September 22-26, 2014.
7. Xu, W., Wei, J., Dolan, J.M., Zhao, H. et al., "A Real-Time Motion Planner with Trajectory Optimization for Autonomous Vehicles," *Robotics and Automation (ICRA), 2012 IEEE International Conference on, IEEE*, 2012, 2061-2067.
8. González, D., Pérez, J., Lattarulo, R., Milanés, V. et al., "Continuous Curvature Planning with Obstacle Avoidance Capabilities in Urban Scenarios," *Intelligent Transportation Systems (ITSC), 2014 IEEE 17th International Conference on, IEEE*, 2014, 1430-1435.
9. Noto, N., Okuda, H., Tazaki, Y., and Suzuki, T., "Steering Assisting System for Obstacle Avoidance Based on Personalized Potential Field," *Intelligent Transportation Systems (ITSC), 2012 15th International IEEE Conference on, IEEE*, 2012, 1702-1707.
10. LaValle, S.M. and Kuffner, J.J. Jr., "Randomized Kinodynamic Planning," *International Journal of Robotics Research* 20, no. 5 (2001): 378-400.
11. Urmson, C., Ragusa, C., Ray, D., Anhalt, J. et al., "A Robust Approach to High-Speed Navigation for Unrehearsed Desert Terrain," *Journal of Field Robotics* 23, no. 8 (2006): 467-508.

12. MacAdam, C.C., "Application of an Optimal Preview Control for Simulation of Closed-Loop Automobile Driving," *IEEE Transactions on Systems, Man, and Cybernetics* 11, no. 6 (1981): 393-399.

13. Kapania, N.R. and Gerdes, J.C., "Design of a Feedback-Feedforward Steering Controller for Accurate Path Tracking and Stability at the Limits of Handling," *Vehicle System Dynamics* 53, no. 12 (2015): 1687-1704.

14. Zhuo, G., Wu, C., and Zhang, F., "Model Predictive Control for Feasible Region of Active Collision Avoidance," SAE Technical Paper 2017-01-0045, 2017, doi:10.4271/2017-01-0045.

15. Nelson, W., "Continuous-Curvature Paths for Autonomous Vehicles," *Robotics and Automation, 1989. Proceedings, 1989 IEEE International Conference on, IEEE*, 1989, 1260-1264.

16. Guo, K. and Ding, H., "Effect of Yaw Moment through Differential Braking under Tire Adhesion Limit," *Automotive Engineering* 24, no. 2 (2002): 101-104.

CHAPTER 5

Basic Autonomous Vehicle Controller Development through Modeling and Simulation

Ayush Goel
IIT Kanpur

Somnath Sengupta
IIT Kharagpur

Autonomous vehicles at various stages will impact the future of transportation by improving reliability, comfort and safety of the passengers. In this paper, for an existing experimental vehicle, fitted with various sensors and actuators typically required by autonomous vehicles, a basic level-1 autonomous controller for braking and throttle actuations is proposed. This controller is primarily developed for stop-and-go scenarios along with the additional functionalities of automatic cruise control (ACC) and automatic emergency braking (AEB). Since the rigorous testing of autonomous vehicle in actual roads can be time consuming, costly and having safety issues, a simulation test-bench based approach is considered to develop and test the controller. The controller, based on practical data is developed in simulation environment to primarily maintain safe distance from surrounding traffic objects while fulfilling requirements such as jerk levels, conditional braking, speed limits, etc. In this work, only a longitudinal controller is developed for low speeds (<30 kmph) and low throttle scenarios for which a four-wheel based vehicle dynamics model is formulated excluding the nonlinear tire model. Experimental data while running the experimental vehicle in actual traffic is acquired from camera, triangulation of ultrasonic sensors, throttle, brake pedal position and velocity of the vehicle and is used for tuning and validation

CITATION: Goel, A. and Sengupta, S., "Basic Autonomous Vehicle Controller Development through Modeling and Simulation," SAE Technical Paper 2018-01-0041, 2018, doi:10.4271/2018-01-0041.

of the derived model, to ensure satisfactory accuracy. Accordingly, a relative distance and relative velocity dependent longitudinal controller comprising of several coordinated PID controllers is designed in stop-and-go scenario and in AEB mode. The captured pre-recorded traffic video along with acquired throttle, braking, speed and relative distance information is synced with the proposed controllers simulation execution for correlations, wherein the acquired relative distance data is used as reference to run the simulations. The proposed practical data based simulation test environment is successful in creating multiple test scenarios and the developed longitudinal controller is able to satisfactorily autonomously control the vehicle in the desired manner.

Introduction

Autonomous vehicles and associated research have been important topics in both industries and academia. They offer us with a possibility of zero-accidents, using 360-degree sensors and efficient computation based decision making that do not get distracted compared to human's frequently interrupted 200-degree field of view.

Work in the autonomous vehicle has grown dramatically over (recently) several years. Though the path to a completely autonomous vehicle crosses several engineering fields that all needs to be integrated into one complete system, the idea of autonomous vehicle can be seen as early as 1920s when only cruise control was adapted in the vehicles [1]. The cruise controller is an example of a longitudinal controller affecting acceleration and deceleration of the car and maintaining a fixed speed. While with a lateral controller we can make the vehicle steer itself, recent developments [2, 3] have accelerated the interest and growth in this area and several car manufacturing companies are doing research in this field. There are 5 levels of autonomous vehicle in which initial 2 levels are basic and major enhancement in technology starts after level 3. Google's self-driving car project, WAYMO has produced autonomous vehicles [4]. Audi has manufactured an autonomous race car R7 [5] and the Mercedes-Benz F 015 luxury in motion autonomous vehicle incorporates many different aspects of day-to-day mobility [6].

The main components of a typical modern autonomous vehicle are [7] a.) Perception which includes cameras and sensors that helps the car sense the dynamic environment, b.) Localization with a high resolution GPS which keeps track of the global position of the vehicle, and c.) Control for motion planning in accordance with the environment and safety regulations. The control of an autonomous vehicle can be further categorized as i.) Longitudinal and ii.) Lateral. To reduce the unnecessary fatigue caused to the driver during a repetitive low-speed start or stop situation (Inching) in traffic, a longitudinal controller is designed to control vehicle's throttle and brake actuations. Moreover, this is implemented to prevent collisions and cover distances quickly from the front object. Autonomous Emergency Braking (AEB) systems which operate during high speeds can be one of the functions of the longitudinal controller which effectively prevents accidents and reduces casualties simultaneously. [8] focuses on the two car scenario in which controller applies brake if the collision is going to happen based on the relative distance from the front object, obtained using a LIDAR or RADAR sensor [9].

To design a controller considering an exhaustive set of driving cases, it is essential to first develop a mathematical vehicle dynamics model to run in simulation environment [10]. Existing schemes reported in literature use a rigid body model [11] which is not accurate with the real behavior of a vehicle. In [12, 13] bicycle model is used, that behaves more closely to a real vehicle but is still a simplification. One can consider using more complex model which is a four wheel model called an extended bicycle model [14] combined with the Pacejka tire models [15].

In this paper, the controller development for level-1 and partially level-2 autonomous vehicle is proposed based on the real field data acquired from an experimental vehicle running in the road. This process, using the real field data is carried out in a simulation environment to minimize the need of extensive set of measurements (under different conditions) and to reduce the potential risks to human lives. The paper begins with showing a comparison between lumped vehicle model and four wheel model and examines how well these two models are able to model a vehicles dynamic behavior compared to real data acquired. Then, results are used to design, develop and validate a longitudinal controller for stop and go scenario common in urban driving under heavy traffic. In the subsequent section, the AEB system in two car scenario [16] is implemented in the same controller for higher speeds. Development of controller is based on relative distance and relative velocity with respect to the front vehicle, which is measured using ultrasonic sensors and camera mounted on the experimental vehicle. The brake and throttle values are also acquired and are used to tune and validate the model under various scenarios.

The proposed approach offers less computational complexity and can be implemented at low vehicle speeds where tire models becomes singular and slip can be neglected. This work presents simulation results which show that the proposed controller provides satisfactory performance compared to the experimental results.

Vehicle Dynamics Models

To represent an actual vehicle in the simulation environment an appropriate mathematical vehicle dynamics model has to be derived. The tire dynamics, if included will make the model computationally expensive. The objective of this work is to make a controller for low speeds, where tire becomes singular. Two mathematical vehicle dynamic models are derived, excluding tire dynamics.

Lumped Vehicle Model

A rigid body model, in Figure 1 is considered with non-deformable wheels, executing pure rolling with no slipping. The kinematic equations for this model is given by equations 1-2,

$$\dot{x} = v \tag{1}$$

$$\dot{v} = A_x \tag{2}$$

Where x is the coordinate of the center of mass and v being its velocity and A_x being its longitudinal acceleration in an inertial frame. The overall dynamic equation given for this model based on Newton's 2nd law of motion is given by equation 3-4,

$$F_T - F_B - F_G - F_F - F_D = MA_x \tag{3}$$

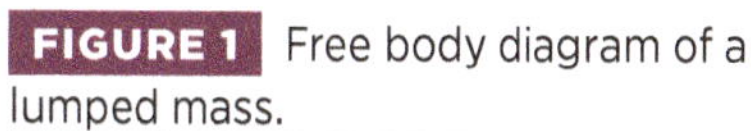
FIGURE 1 Free body diagram of a lumped mass.

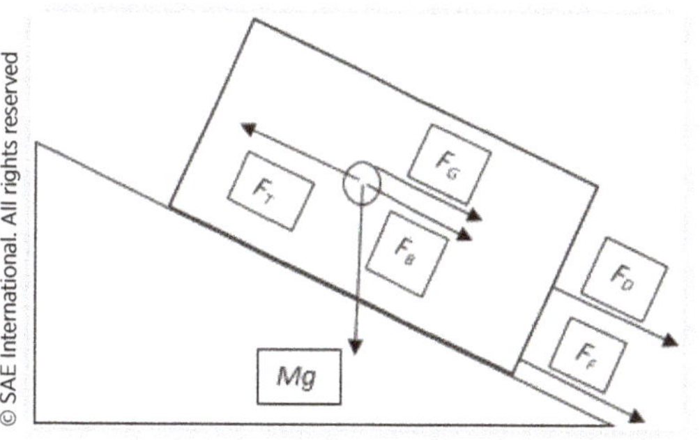

$$F_T - F_B - 1/2\rho v^2 AC_d - \mu Mg\cos\theta - Mg\sin\theta = MA_x \tag{4}$$

Where,
F_T - Tractive Force
F_B - Braking Force
A - Frontal Area of the vehicle
C_d - Drag coefficient
μ - Coefficient of Friction
M - Mass of the Vehicle
g - Acceleration due to Gravity
θ - Angle of the gradient

Four Wheel Model

In this model development, as shown in Figure 2 the approach is to derive full equations of motion with as few approximations as possible. In order to gain a better understanding, both front and rear wheels dynamics are studied separately to develop the model for the longitudinal motion. These dynamics interact with engine torque, braking torque, aerodynamic force and wheel rotational dynamics.

$$W_f = \frac{MgL_r}{L} - \frac{MA_xH}{L} \quad M_f = \frac{W_f}{2g} \tag{5}$$

$$W_r = \frac{MgL_f}{L} + \frac{MA_xH}{L} \quad M_r = \frac{W_r}{2g} \tag{6}$$

The longitudinal weight transfer is studied first, using Newton's second law on an accelerated vehicle and the load supported by each wheel [17] is given by equations (5-6),

The wheels are assumed to have inertia I, and dynamic equations are derived considering pure rolling and no slipping between ground and tire. The differential is assumed to be a lock differential and same value of coefficient of friction and coefficient of rolling resistance is assumed for all tires.

The simplified dynamic equation for each wheel is derived and then all are assumed to have same longitudinal acceleration without slipping and hence the substituted equations for full vehicle are captured by equations (7-8),

<u>During Acceleration</u>

$$2\frac{\tau_e}{R}\left[\frac{(M_rR^2 - I)}{(M_rR^2 + I)}\right] - \frac{1}{2}\rho v^2 C_d A - Mg\sin\theta - CrrMg = MA_x \tag{7}$$

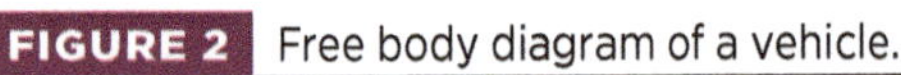
FIGURE 2 Free body diagram of a vehicle.

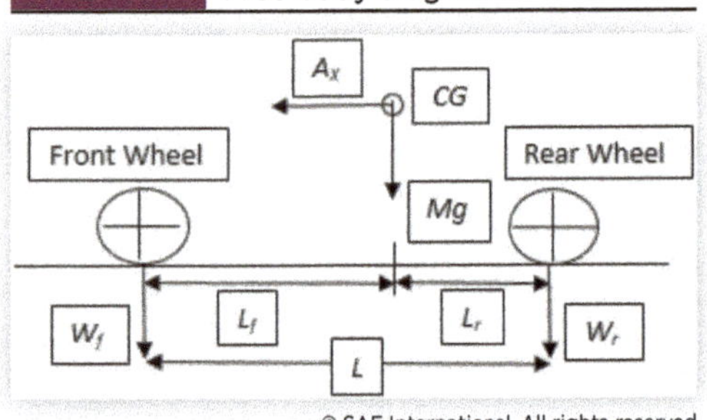

<u>During Deceleration</u>

$$-\tau_B\left[\frac{MR}{(I + M_rR^2)}\right] - \frac{1}{2}\rho v^2 C_d A - Mg\sin\theta - CrrMg = MA_x \tag{8}$$

Where,
τ_e - Engine Torque
τ_b - Braking Torque
W_f - Normal reaction on the front wheels
W_r - Normal Reaction on the rear wheels
L - Wheelbase of the vehicle

L_f - Distance of CG from the front wheels
L_r - Distance of CG from the rear wheels
A_x - Longitudinal acceleration of the vehicle
H - Height of CG above ground
Crr - Coefficient of rolling resistance
I - Inertia of a single wheel

These equations give direct correlation between vehicle's acceleration and torque produced by the engine similarly, between deceleration and the braking torque generated on the brake rotors. Hence, the equations are used to govern the throttle and brake commands to go from one state of motion to other state.

Controller Development in Simulation Test-Bench

The longitudinal controller is developed in various modes for different required cases in the simulation environment as shown in Figure 3. The controller uses the relative distance, relative velocity (from the front vehicle) and vehicles absolute self-velocity (governing parameters) as its inputs and uses PID controllers to accordingly give throttle and brakes commands as outputs. These outputs are consequently used in plant model to get final output of vehicle's velocity.

The inching mode activates when the distance from the front object is <120 inches and the relative velocity is negative in sign (front vehicle velocity is < vehicle self velocity). In this mode the controller maintains the vehicle at a speed <15 kmph and avoids sudden braking to provide slow deceleration and minimize the jerk while keeping a minimum distance of 12 inches with the front vehicle. An extension to this is a cruise mode which activates when the relative velocity is positive or zero and it maintains the vehicle at constant 15 kmph speed.

The non-inching mode activates when the distance is >200 inches from the front object. A set controller algorithm calculates the suitable speed of the vehicle governed by relative distance and velocity and maintains the vehicle at calculated speed by giving suitable brake and throttle actuations. AEB mode, an extension to non-inching mode activates when the vehicle's velocity is >30 kmph and the relative distance decrease suddenly to 120 inches. In this mode the controller applies hard braking giving a maximum deceleration of 6 ms^{-2}.

The controller is validated using some simple experiments done in different traffic conditions on road, which is shown in results and analysis section. To validate the

FIGURE 3 Different modes of longitudinal controller.

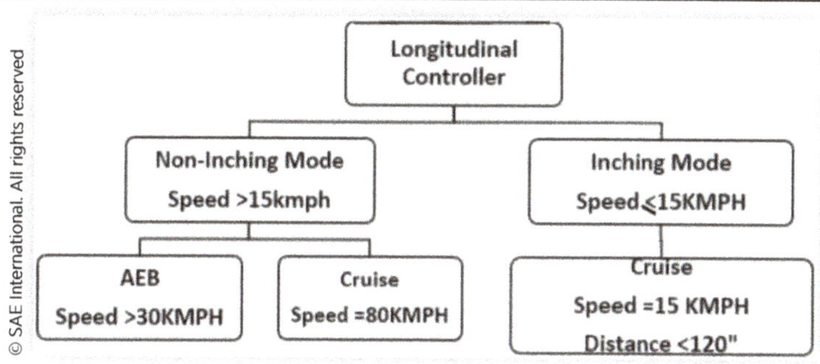

FIGURE 4 Closed loop feedback system for controller in simulation environment.

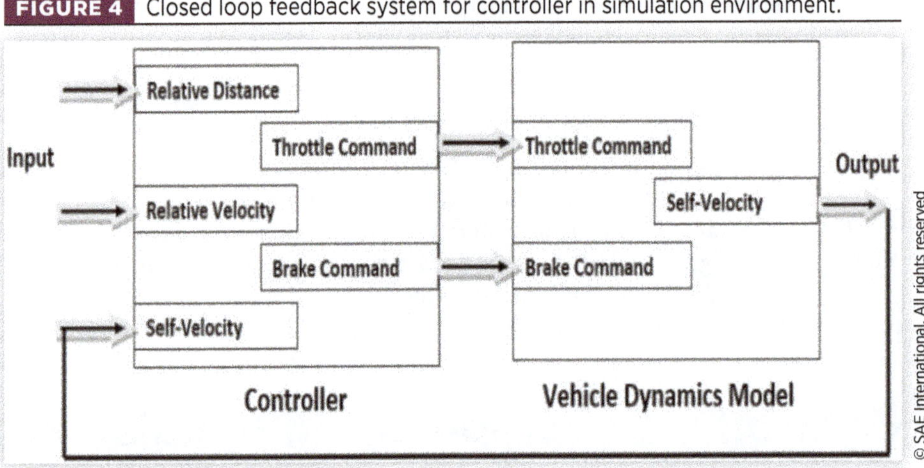

controller in numerous complex scenarios a simulation test-bench is further created in Mathworks Simulink environment, which takes the input of front vehicle's acceleration signal and self vehicle velocity to compute relative distance and relative velocity, which is fed into a closed loop feedback system of the controller depicted by Figure 4.

Experimental Vehicle Setup

The experiments are performed on a running Hyundai i10 vehicle [19], acquiring data of brake pedal position and corresponding vehicle speed at a uniform sample time of 200 ms. The experimental vehicle under consideration is equipped with 4 ultrasonic sensors, each having a span of 70 degrees and each mounted at the front deck, a camera of span 140 degrees mounted at front and proximity sensor for brake pedal position and vehicle speed sensor for measuring speed of the vehicle. For low speeds and stop-and-go mode the initial throttle position is set fixed to 14%. However, accounting for vibrations in the vehicle we can assume it to be varying in a small range of 11%-18%. The ultrasonic sensors are used to measure the distance between the vehicle and front object.

Working of Ultrasonic Sensors

The ultrasonic sensor [18] is used to measure the relative distance of an object in front. The sensors transmit and receive the ultrasonic signal and wave propagation takes place at the speed of sound. It takes approximately 0.3 ms for the sound wave to propagate through 1 m distance. Each sensor gives the output of 3 pulses, distance of the 3 nearest points of object to itself.

Sensor Arrangements and Interpretation

The 4 Ultrasonic sensors are positioned in a way that, 2 of the sensors are mounted to sense the object in the direction of motion of the vehicle and 2 for the sideways objects. All 4 sensors combined theoretically gives a total span of 280 degrees. However, effectively, the sensing spans only 180 degrees, accounting for the overlapping regions. The triangulation based fusion is applied to identify the exact location of the object [16].

Drawbacks

1. The ultrasonic sensors are only able to give position of the nearest point which is not enough to get the dimensions of the front object.
2. The sensors located sideways create confusion in the controller as, those objects which are not in the line of motion do not have much significance comparable to the object in the direction of motion.
3. The ultrasonic sensors are not very accurate for large distances as for a distance of only 200 cm they introduce a lag of 10 ms in the measurement. So, beyond 200 cm ultrasonic sensors are not recommended because of decrease in accuracy and also the signal becomes weak as the object goes further and further.

Tuning and Model Validation

The simulation environments are setup for both the models, to investigate how well these developed models are able to predict the vehicles state compared to the real data acquired from the running vehicle in the road which is described in results and analysis section. The plant has inputs of throttle and brake pedal position giving an output of vehicle's estimated velocity. Data acquired from an experimental vehicle only gives information about the vehicle's velocity and brake pedal position. In the absence of throttle position, it can be predicted in the simulation environment to make the resultant vehicle dynamics behave like that of an actual vehicle. Tuning the model will facilitate with the values of some immeasurable quantities which will further improve the accuracy of model. An approximate engine map is generated for low speeds (<30 kmph) and low throttle range (11%-18%). To estimate the engine map a PID controller is used to generate desired torque from the engine, based on the velocity and brake command. Then a non-linear map is derived for that throttle range (11%-18%).

Subsequently, PID controller is used to predict the throttle pedal position. Both throttle (derived) and brake position (from the data) is then used as input to the plant to get the modeled vehicle's velocity. Moreover, this velocity is later compared with the actually measured velocity from the data.

To understand the performance of model, average percentage error between real and model velocity is calculated and is shown in results and analysis section.

Results and Analysis

The controller is tested in various vehicle operating modes, discussed earlier in controller development section, using real data acquired under those conditions, creating different sets of combinations of operating conditions in the simulation test-bench environment. First, the model validation for the real speed tracking based on the real acquired data is done. Then the controller is tested in the stop-and-go low speed mode and then the real acquired data and traffic video from experimental vehicle is synced to run simulations which are done in MATLAB Simulink using computer vision toolbox. Model's velocity is then calculated and compared with real vehicle's acquired velocity.

Model Validation

The real data acquired by experimental vehicle on road, running with constant throttle and acquiring values for brake command Figure 5 and corresponding vehicle self velocity is used to compare two vehicle dynamic models derived earlier in this paper. The model which is able to follow the velocity trend more closely is elected for implementing the simulation test bench.

The lumped vehicle dynamics model in Figure 6, has an average percentage error of 16.12% in predicting the state of the vehicle, while the four wheel model in Figure 7, has average percentage error of 9.3% between the predicted and the measured value of velocity of the vehicle. The four wheel model is able to predict the vehicle's state more accurately than the lumped model, as depicted by their respective results. Therefore, throughout the controller development process, the four wheel model is used.

FIGURE 5 Experimentally acquired brake command.

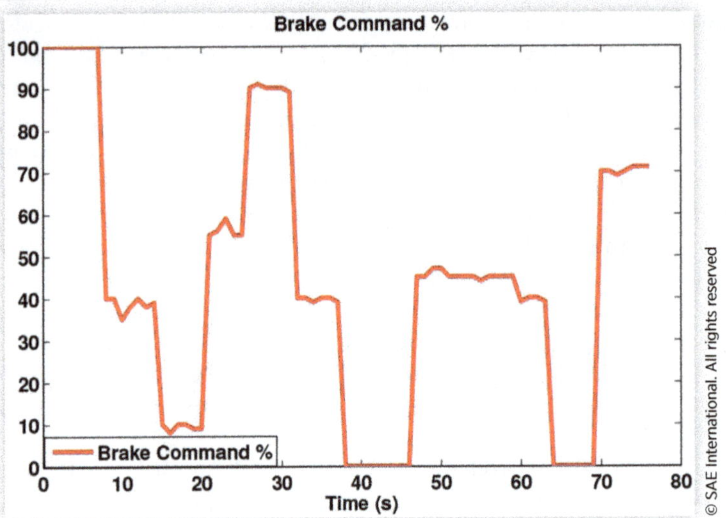

FIGURE 6 Lumped vehicle dynamics model.

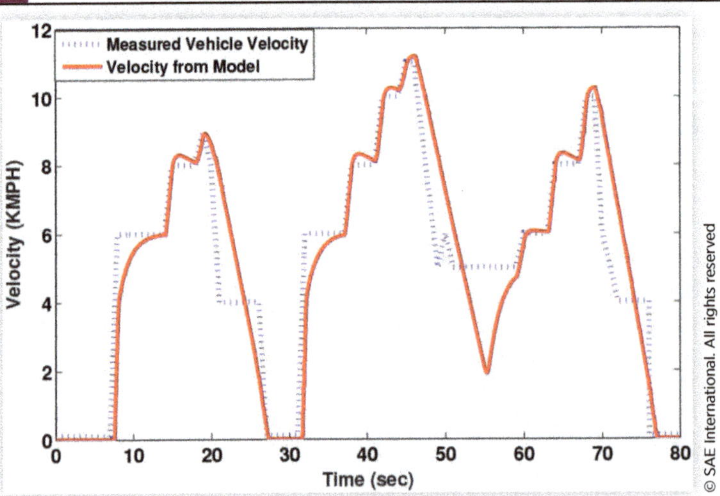

FIGURE 7 Four wheel vehicle dynamics model.

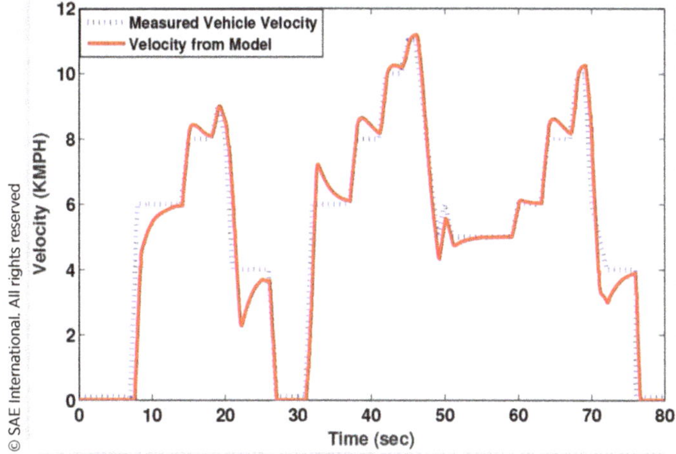

Inching Mode

The stop-and-go scenario common in traffic jams, in urban driving is referred as inching mode in this paper. To provide safe and comfortable ride to the passengers certain aspects are considered as requirements such as, a. Small jerk levels, b. Cushion braking, c. Smooth transition between different modes, d. Maintaining a minimum safe distance at all times and e. Minimizing the throttle while braking for better control.

The simulation shown in Figure 8 is carried out in simulation test-bench and controller measures the output velocity based on the relative distance and velocity from the front object. To attain this velocity the controller provides brake and throttle

FIGURE 8 Test-bench performance in inching mode.

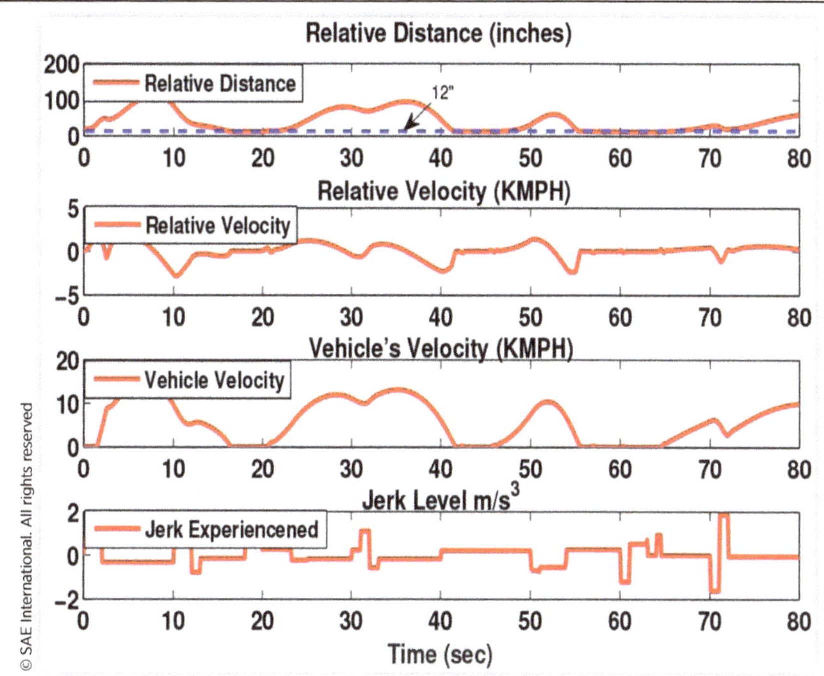

CHAPTER 5

actuations using closed loop PID controller, which are fed to vehicle dynamics model. The vehicle dynamics model respond to those actuations and output the desired velocity.

The vehicle velocity goes to zero as the relative distance grazes the minimum 12 inches bar. The controller is able to make vehicle follow the expected trend, i.e. increase in speed when relative distance or relative velocity increases and vice-versa. The jerk levels are maintained in the small range of −2 to +2 ms-3 which is evident from Figure 8.

Cruise Mode

This mode is the extension to inching mode and it activates when the relative distance reaches 120 inches and deactivates when relative distance is >200 inches (done by using a hysteris based control). This mode is required to provide a smooth transition between inching and non-inching mode. The controller in this mode tends to operate the vehicle at constant speed of 15 kmph.

Figure 9 shows the variation of relative velocity and relative distance for some given front vehicle's acceleration, based on which controller's throttle and brake command are also shown. In Figure 10, the acquired velocity of the vehicle is plotted against time. It can be observed that the vehicle is in cruising range and the controller is able to keep vehicle's velocity constant to 15 kmph and afterwards it acts accordingly with relative distance and velocity.

Real Data Validation with Video

In this case the experimentally acquired real data while running on road is compared with the running controller with all cases including cruise, AEB, Inching, Non-inching and this is correlated with the corresponding video. The video is setup in sync with the controller output to visualize the controller performance in the real scenario, for this MATHWORKS environment and it's computer vision toolbox is used.

FIGURE 9 Cruising mode parameters.

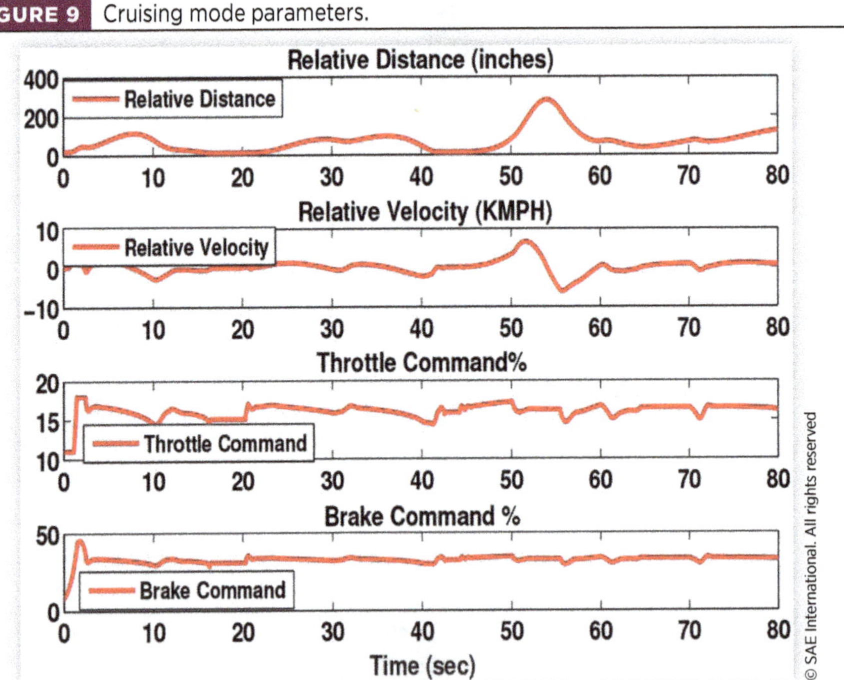

FIGURE 10 Test-bench performance in cruising mode.

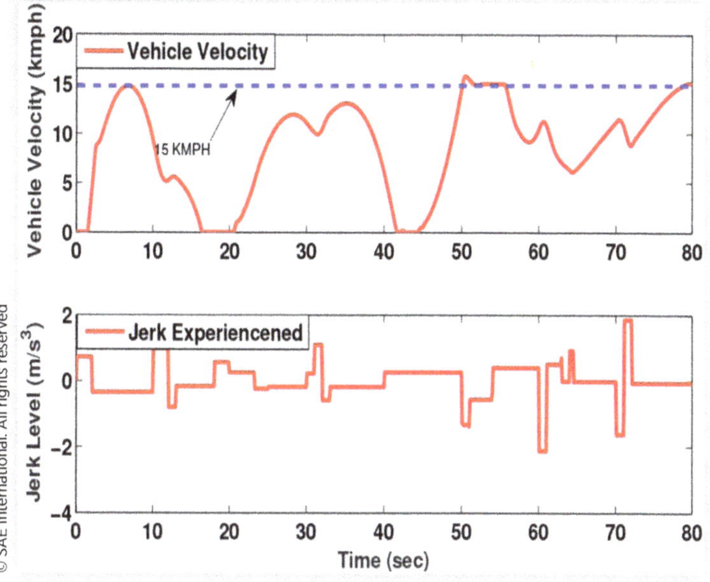

TABLE 1 Real data time-stamp

Real Data Time-Stamp					
Time-Stamp	**Sensor 1 (cm)**	**Sensor 2 (cm)**	**Sensor 3 (cm)**	**Sensor 4 (cm)**	**Actual Speed (KMPH)**
10:37:20	200	200	200	70	24
10:37:20	200	200	200	70	24
10:37:20	200	200	200	70	24
10:37:20	200	200	200	70	24
10:37:20	200	200	200	70	22

Table 1 shows from a logged data at an instance of time when the experimental vehicle is driven on a straight road longitudinally and its action is observed in the real time, with moving objects in front. The data displayed is from the ultrasonic sensors of span 70 degrees mounted at the front deck of the experimental vehicle sampling at 200 ms.

In Figure 12 a front vehicle is marked and time stamp at which the image is taken along with with the vehicle's self velocity is also shown. The image is taken from the video running in sync with the controller in real time. The step-size for the discrete controller is 0.416666 s same as that for the video captured from the camera having a span of 140 degrees mounted in front of the experimental vehicle. The distance of this vehicle is observed using the Table 1, value from 4th sensor is neglected in which only one object is present in front, so the distance is 200 cm (80 inches) and the actual self vehicle's velocity is also observed to be 22 kmph.

In Figure 11 the comparison between real velocity acquired on the road in traffic condition and model performance based on acquired relative distance value is done. It is observed that the simulated model velocity is successfully able to follow the real experimental velocity trend. The model is successful because it is responding to the change in relative distance and relative velocity in the same way as driver does and that is directly imparted in the trend of vehicle's velocity. Furthermore, the root mean square error for speeds <25 kmph is found out to be 1.17 kmph which is acceptable, given in

FIGURE 11 Comparison of real time data with model performance.

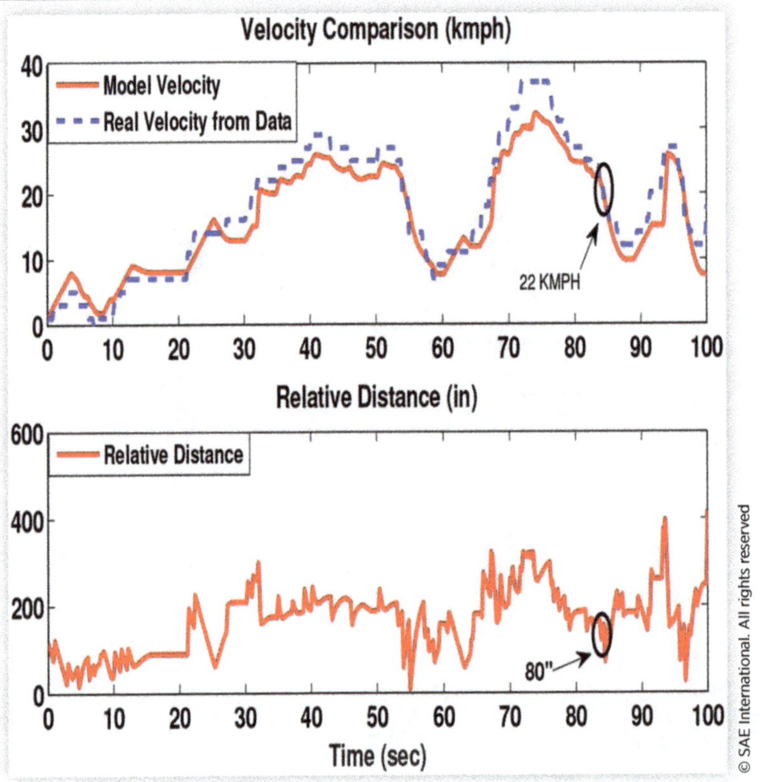

FIGURE 12 Image at an instance from real captured video on road.

the different operating conditions such as different air density, different inflation pressure in tires, slope of the road, etc. between real environment and simulation environment. The controller is giving satisfactory results for low speeds but as speed increases further the model tends to lag, probably due to tire dynamics incorporated into the equation which is considered singular to decrease the complexity of the model.

Figure 11 is showing the showing the instance of 22 kmph and 80 inches relative distance at the corresponding time stamp according Table 1 and Figure 12 shows the snapshot from the real traffic video at same corresponding time stamp. At this instant it shows the model velocity is approximately same as the real acquired velocity.

Conclusion

The controller for various operating modes of a level 1 autonomous vehicle is developed. A universal mathematical vehicle dynamics simulation model is derived which is validated with experimental data and can be applied to any vehicle upon changing suitable inertia parameters. A controller is developed in the low speed stop-and-go mode which is able to maintain a minimum distance which can be set as per the requirement and is able to reduce the brake while accelerating to reduce the fuel consumption. The non inching mode in the controller can be modified and improved if more experimental data such as engine map and throttle data in higher speeds from testing are available. The proposed approach offers very less complexity and can create any case in the simulation itself. The present simulation results have shown that the proposed controller provides satisfactory performance compared to the experimental results. In future the tire dynamics will be implemented in vehicle dynamics models to ensure controller development and validation at high speeds. Thereafter the coupled longitudinal and lateral controller can be incorporated to implement level-3 autonomous vehicle level functionality.

Acknowledgement

The present research work is supported by the RnD unit of KPIT Technologies Ltd. The authors are grateful to KPIT's Autonomous Vehicle group and help from Mr. Rahul Uplap, Associate Vice President for his helpful suggestions and providing directions for setting up requirements and performance expectations from the autonomous controller and its validation.

Contact Information

Ayush Goel, Masters in Engineering
Department of Aerospace Engineering, IIT Kanpur
Phone: +91-7755057617
ayushg46@gmail.com

Somnath Sengupta, PhD
Assistant Professor, IIT Kharagpur
Phone: +91-9604751288
somnath.el21@gmail.com

References

1. Lafrance, A., "Your Grandmother's Driverless Car," *The Atlantic* (June 29, 2016).
2. SAE International, "Auto-Drive Challenge," accessed May 2017, http://students.sae.org/cds/autodrive/event/.
3. Thrun, S., Montemerlo, M., Dahlkamp, H., Stavens, D. et al., "Stanley: The Robot that Won DARPA Grand Challenge," *Journal of Field Robotics*, doi:10.1002/rob.20147.
4. Reily, M., "Waymo Has Invited the Public to Hop into Its Self-Driving Cars," *MIT Technology Review* (April 25, 2017).
5. Ramey, J., "Watch an Autonomous Audi RS7 Fly around the Hockenheim Circuit," *Autoweek* (October 21, 2014).
6. Oagana, A., "2015 Mercedes-Benz F-015 Luxury in Motion," *Topspeed Cars* (January 6, 2015).
7. Ziegler, J., Dang, T., Franke, U., Lategahn, H. et al., "Making Bertha Drive-An Autonomous Journey on a Historic Route", *IEEE Intelligent Transportation Systems Magazine* 6, no. 2 (2014): 8-20, ISSN:1939-1390, doi:10.1109/MITS.2014.2306552.
8. Doecke, S.D., Anderson, R.W.G., Mackenzie, J.R.R., and Ponte, G., "The Potential of Autonomous Emergency Braking Systems to Mitigate Passenger Vehicle Crashes," *Australasian Road Safety Research, Policing and Education Conference*, Wellington, October 4-6, 2012.
9. Lee, I. and Luan, B., "Design of Autonomous Emergency Braking System Based on Impedance Control for 3-Car Driving Scenario," SAE Technical Paper 2016-01-1453, 2016, doi:10.4271/2016-01-1453.
10. Kong, J., Pfeiffer, M., Schildbach, G., and Borreli, F., "Kinematic and Dynamic Vehicle Models for Autonomous Driving Control Design," *Intelligent Vehicles Symposium (IV), 2015 IEEE*, ISSN:1931-0587, doi:10.1109/IVS.2015.7225830.
11. Kuhne, F., Manoel, J., and Lages, W.F., "Mobile Robot Trajectory Tracking Using Model Predictive Control," VII SBAI/II IEEE LARS, So Lus, setembro de 2005.
12. Falcone, P., Borrelli, F., Asgari, J. et al., "Predictive Active Steering Control for Autonomous Vehicle Systems," *IEEE Transactions on Control System Technology* 15, no. 3 (May 2007), doi:10.1109/TCST.2007.894653.
13. Keviczky, T., Falcone, P., and Borrelli, F., "Predictive Control Approach to Autonomous Vehicle Steering," *American Control Conference, IEEE*, ISSN: 2378-5861, doi:10.1109/ACC.2006.1657458.
14. Turri, V., Carvalho, A., Tseng, H.E. et al., "Linear Model Predictive Control for Lane Keeping and Obstacle Avoidance on Low Curvature Roads," *Intelligent Transportation Systems-(ITSC), 2013 16th International IEEE Conference on*, ISSN:2153-0017, doi:10.1109/ITSC.2013.6728261.
15. Woods, R.L., "Normalization of the Pacejka Tire Model," SAE Technical Paper 2004-01-3528, 2004, doi:10.4271/2004-01-3528.
16. Wijk, O., Jensfelt, P., and Christensen, H.I., "Triangulation Based Fusion of Ultrasonic Sensor Data," *Robotics and Automation, 1998. Proceedings. 1998 IEEE International Conference on*, ISSN:1050-4729, doi:10.1109/ROBOT.1998.680966.
17. Milliken Jr., W.F. and Milliken, D.L., *Race Car Vehicle Dynamics*, (Warrandale: SAE International, 2003), ISBN:1-56091-526-9.

18. Girardi, W.J., "Limitations of Ultrasonic Obstacle Sensors for Industrial Lift Truck Applications," SAE Technical Paper 961809, 1996, doi:10.4271/961809.

19. "Hyundai i10 Vehicle," accessed May 2017, http://www.hyundai.com/in/en/Showroom/Cars/i10/PIP/index.html.

20. Dadras, S., "Path Tracking Using Fractional Order Extremum Seeking Controller for Autonomous Ground Vehicle," SAE Technical Paper 2017-01-0094, 2017, doi:10.4271/2017-01-0094.

CHAPTER 6

A Study of Automatic Allocation of Automotive Safety Requirements in Two Modes: Components and Failure Modes

David Parker and Yiannis Papadopoulos
University of Hull

Antoine Godof and Laurent Saintis
University of Angers

ISO 26262 describes a safety engineering approach in which the safety of a system is considered from the early stages of design through a process of elicitation and allocation of system safety requirements. These are expressed as automotive safety integrity levels (ASILs) at system level and are then progressively allocated to subsystems and components of the system architecture. In recent work, we have demonstrated that this process can be automated using a novel combination of model-based safety analysis and optimization metaheuristics. The approach has been implemented in the HiP-HOPS tool, and it leads to optimal economic decisions on component ASILs. In this paper, first, we discuss this earlier work and demonstrate automatic ASIL decomposition on an automotive example. Secondly, we describe an experiment where we applied two different modes of ASIL decomposition. In HiP-HOPS, it is possible to decompose ASILs either to the safety requirements of components or individual failure modes of components. Protection against independent failure modes could, in theory, be achieved at different ASILs and this will lead to reduced design costs. Although ISO26262 does not explicitly support this option, we have studied the implications of this more refined decomposition on system costs but also on the performance of the decomposition process itself, and we report on the results. Finally, motivated by our study on ASIL

CITATION: Parker, D., Papadopoulos, Y., Godof, A., and Saintis, L., "A Study of Automatic Allocation of Automotive Safety Requirements in Two Modes: Components and Failure Modes," SAE Technical Paper 2018-01-1076, 2018, doi:10.4271/2018-01-1076.

decomposition, we discuss the general need for increased automation of safety analysis in complex systems, especially autonomous systems where an infinity of possible operational states and configurations makes manual analysis infeasible.

Introduction

Systems of classification for different levels of safety integrity have been introduced in several different safety standards. While the safety standard IEC 61508 first popularized the Safety Integrity Level (SIL), other safety standards such as ISO 26262 and ARP4754-A developed domain specific versions. The aerospace industry, for example, defines the Development Assurance Level (DAL) in their ARP4754-A standard. ISO 26262, an automotive safety standard [1] defines the Automotive Safety Integrity Level (ASIL) which is the focus of the work in this paper, though the principles are applicable generally across domains.

One of the purposes of the ASIL is to address the issue of traceability with regards to safety in the design of systems. This should be applicable from the early stages of the design process, while initial concepts are being considered, right through to the operational phases of the final product and capture how requirements have been refined and met by the design.

The inevitable and increasing use of software systems in place of purely mechanical systems has meant that traditional techniques of expressing safety requirements as maximum target probabilities for system failures are no longer sufficient.

The ASIL concept is used instead to represent the stringency of safety requirements with respect to software and systematic failures in general. They range from ASIL A (least strict) to ASIL D (most strict). Additionally, QM is used when no special safety requirements are needed indicating only routine Quality Management should be applied.

The elicitation of these safety requirements, as prescribed by the ISO 26262 standard, begins with a hazard and risk analysis to identify potential malfunctions and their hazardous consequences. Based on the severity, likelihood, and controllability of the identified hazards an ASIL is assigned to the hazard to generate the necessary requirements to ensure that any associated risks are reduced to an acceptable level.

Traceability is partially delivered through the process of allocation and decomposition of the ASILs throughout the sub-systems and sub-functions of the system as it is refined from the early concepts. The ISO 26262 standard describes how components that directly cause a hazard receive the ASIL of the hazard. It also lays out guidelines for where multiple components must be involved to cause the hazard. In this instance the components can share the burden of complying with the hazard's ASIL. A process of decomposition (described further later) is defined by the standard to specify what options are allowed when distributing the load of responsibility for meeting a hazards ASIL.

However, the practical application of this decomposition is fraught with difficulty. It requires practitioners to have intricate knowledge of the system being considered including the consequences of architectural failure behavior and how it propagates through the system. This problem is exacerbated by the increases in complexity found in modern systems with more and more interconnected functions delivered through a mix of software and hardware. An explosion of possible operational states, particularly in autonomous systems that are required to work in heterogeneous environments make it even more difficult. The lack of supporting examples in the ISO 26262 standard is not helpful here and the lack of clarity can often lead to mistakes [2].

A further consideration that is not provided by the standard is that meeting the safety requirement is not the end of the story when it comes to the practical application of the guidance. Coming up with a decomposition of ASILs in a system that satisfies the safety requirements of the identified hazards is a difficult task by itself. However, doing so is in fact merely meeting a constraint. Once that constraint has been met (or in meeting that constraint) it becomes necessary to consider the cost implications of doing so.

Applying different levels of stringency to the safety processes of system development has knock on effects on the cost of said development. The ability to allocate and decompose ASILs in a system in a cost effective (even cost optimal) way further strengthens the need for automated methods.

Various approaches have been made to provide automated assistance to the problem of ASIL decomposition beginning with an exhaustive deterministic method [3], and including optimization approaches such as linear programming [4], exact solvers [5], penalty-based genetic algorithms [6], and Tabu-search [7].

The remainder of the paper will outline a case study that will be used to illustrate the process of modelling a system for ASIL decomposition. It will highlight the need for an automated process for applying the decomposition in a cost optimal way and how to do this using a variation on earlier work [7]. Finally, it will discuss the results of applying the process at different levels of granularity (components versus their failure modes) and the implications of doing so.

Hybrid Braking System Case Study

The effects of the different decomposition techniques will be illustrated on the following example system (in more detail [8, 9]) shown in Figure 1. It is designated a 'hybrid'

FIGURE 1 The hybrid-braking system example.

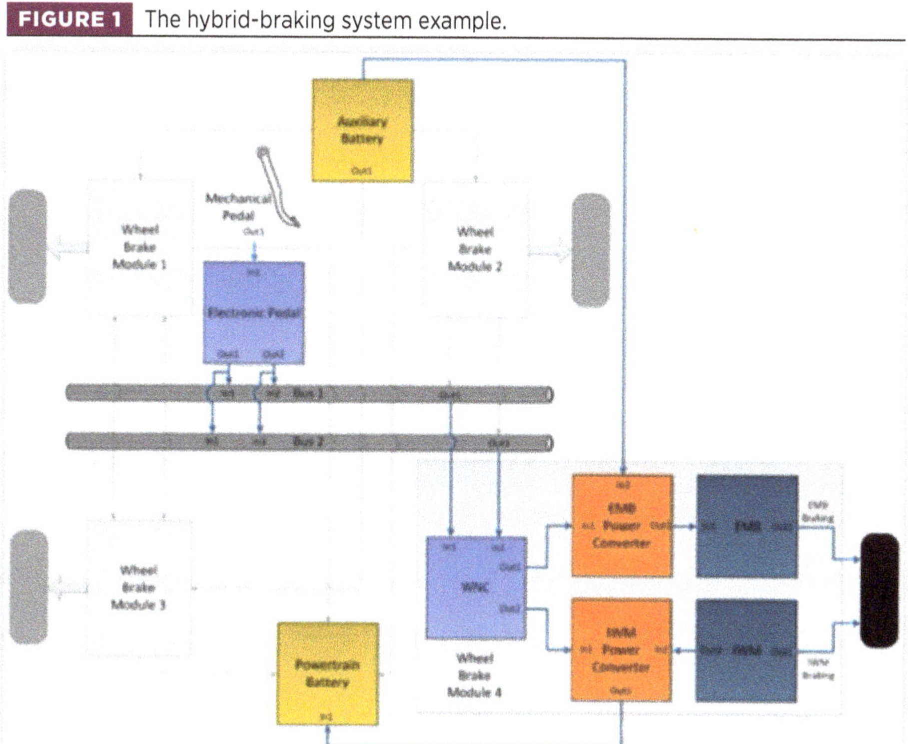

TABLE 1 This table shows the assigned ASILs for the top-level hazards of the system

#	Hazard	ASIL
1	Incorrect Value Braking	D
2	Loss of Braking Rear Wheels	C
3	Loss of Braking Front Wheels	D
4	Loss of Braking Diagonal Wheels	C
5	Loss of Braking 3 out of 4 Wheels	D
6	Loss of Braking All Wheels	D

braking system as the braking effort is provided through the combination of electro-mechanical brakes (EMB) and the regenerative energy capture from the in-wheel motors (IWM).

Driver intention is delivered through a mechanical pedal that is sensed and processed through an electronic pedal unit in this brake-by-wire system. The system comprises 4 wheel braking modules, each able to operate independently. In the diagram, wheel brake module 4 shows detail of its components that is matched but not displayed by the other 3 modules. Braking instructions delivered through the redundant duplex communications bus are received by the wheel node controllers (WNC). The WNCs calculate the action required from the wheel's EMB and IWM actuators and deliver the instructions to the respective power converters. The IWM can provide braking functionality by converting the kinetic energy of the vehicle to electric charge which is delivered to the main powertrain battery. This has the benefit of increasing the range of the vehicle, but at high speeds and periods when the battery is in a full state of charge the full braking needs of the vehicle cannot be met. Hence the need for the partnering EMB. The EMB draws power from an auxiliary battery.

In this example, the hazards in Table 1 were identified for the system and, based on the severity of the hazard, the respective ASILs were assigned to them.

The first hazard in the table represents the scenario where a particular value of braking is requested from the system and a different value is delivered. This could result in either excessive or insufficient braking. The remaining hazards in the table represent an omission of braking (i.e. braking is requested and none is delivered) in combinations of one or more of the 4 wheels.

System Modelling

An important part of the methodology being used here is the ability to iterate on the design. To that end, all of the information being used in the process is derived directly from the system model and provides traceability back from the results to the original model.

The topology of the system model has been modelled in Matlab and Simulink. It is provided by the components, their port interfaces, and the connections between them.

The system's failure model is provided by augmenting the topological model with local failure behavior for each of the components. This local failure behavior is added to the model using HiP-HOPS failure expressions. They describe how deviations of output in a component are caused by either an internal failure of the component or through the propagation of failure from elsewhere in the model represented as a deviation of input of the component.

For example, in this case study model, the EMB Power Converter can fail with an omission of output. This can be caused either by an internal omission causing failure (OFailure) or by an omission deviation of either of its two inputs.

Omission-Out = Omission-In1 or Omission-In2 or OFailure.

In contrast the WNC component has two outputs. Each of them can fail by omission, but this deviation of output is either caused by a specific internal failure (OFailure1 and OFailure2 respectively) or by the combination of an omission deviation at both of the inputs.

Omission-Out1 = (Omission-In1 and Omission-In2) or OFailure1.

Omission-Out2 = (Omission-In1 and Omission-In2) or OFailure2.

Note that at this stage, the system is under design so the precise internal electrical/mechanical/functional component failures are not known. However, the design intention is known and therefore what constitutes potential output failures and their intended relationship to input failures is known. Beyond this, one can hypothesise that each output failure can be caused by one yet unspecified collective internal cause. It is precisely these requirements for avoidance and containment of these internal causes that the decomposition exercise tries to establish via analysis of propagation and effects of those causes of failure. Each failure expression describes a mini-fault tree and each of the components in the system may have one or more to describe how the component propagates, generates, or mitigates failure that it is presented with.

Any deviations of output are propagated through the connections in the model to the inputs of the connected components. In the example, the first output of the WNC is connected to one of the inputs of the EMB Power Converter. Matching failure classes (e.g. Omission) found at either end of such a connection allow the mini-fault trees to be joined.

For example, the omission of the second input of the EMB Power Converter can be replaced by the expression for the omission of the first output of the WNC.

Omission-EMBPC.Out = Omission-EMBPC.In1 or **(Omission-WNC.In1 and Omission-WNC.In2) or WNC. OFailure1** or EMBPC.OFailure.

The part of the expression that relates to the WNC is shown above in bold and additional identifiers have been added to indicate which component the ports and failures originate in.

This process of combining the mini-fault trees of the components begins at the hazards that have been identified for the system. These are connected to the outputs of the systems using the same Boolean expressions. For example, the hazard "Loss of braking of all wheels" is connected using the following expression:

Omission-Brake_Unit1.Braking and.
Omission-Brake_Unit2.Braking and.
Omission-Brake_Unit3.Braking and.
Omission-Brake_Unit4.Braking.

Each braking units' omission of output as a failure expression that refers to an omission of both the EMB and the IWM function, and so on. This process of combining the mini-fault trees of the components in the model continue until all of the connected input deviations have been resolved.

The result is a complete fault tree that is generated for each hazard defined for the system. The fault tree describes the propagation of failure from the internal failures of the components (the basic events are the leaf nodes of the tree) to the top-level hazards of the system through the combination of Boolean logic.

To be used for the ASIL decomposition process it is necessary to have the fault propagation in it minimal form. This is provided through the automatic fault tree analysis capabilities of the HiP-HOPS engine and results in a set of minimal, non-redundant, cut sets.

For the case study example this results in 6 fault trees (one for each of the hazards), each of which shares branches with the others. Consequently, the cut sets that are generated as the result of the fault tree analyses will be shared across multiple hazards. Table 2 shows the number of minimal cut sets generated for each of the hazards.

The cut sets are important for the ASIL decomposition process as each minimal cut set gives a combination of failure

TABLE 2 This table shows the number of minimal cut sets for each of the top-level hazards of the system

#	Hazard	Cut Sets
1	Incorrect Value Braking	1302
2	Loss of Braking Rear Wheels	103
3	Loss of Braking Front Wheels	103
4	Loss of Braking Diagonal Wheels	202
5	Loss of Braking 3 out of 4 Wheels	3136
6	Loss of Braking All Wheels	6727
	total	11573

TABLE 3 This table shows the algebraic value for each ASIL

ASIL	Algebra Value
QM	0
A	1
B	2
C	3
D	4

TABLE 4 This table shows the ASIL algebra of possible choices for decomposition due to the powertrain and auxiliary battery cut set for the "loss of braking rear wheels" hazard. The shaded area shows the configurations that exactly meet the requirement

Powertrain battery		Auxiliary battery		
ASIL	algebra	ASIL	algebra	sum
D	4	D	4	8
D	4	C	3	7
C	3	D	4	7
D	4	B	2	6
B	2	D	4	6
C	3	C	3	6
D	4	A	1	5
A	1	D	4	5
C	3	B	2	5
B	2	C	3	5
D	4	QM	0	4
QM	0	D	4	4
C	3	A	1	4
A	1	C	4	4
B	2	B	2	4
A	1	B	2	3
B	2	A	1	3
C	3	QM	0	3
QM	0	C	3	3
A	1	A	1	2
B	2	QM	0	2
QM	0	B	2	2
A	1	QM	0	1
QM	0	A	1	1
QM	0	QM	0	0

modes that is both necessary and sufficient to cause the hazard. For example, one of the cut sets of the "Loss of Braking Rear Wheels" hazard is an internal omission causing failure of both the auxiliary battery and the powertrain battery.

In particular, the cut sets of order 2 or more (non single points of failure) derived directly from the model show the subsystem independence that is required for decomposition.

The ASIL that has been assigned to this hazard is C. In order to satisfy the safety requirements of the system for this cut set, the ASIL of each of the failures in this cut set could be developed to ASIL C also. However, the ASIL decomposition described in ISO 26262 allows for the allocation of reduced stringency where independent redundancy can be shown. In this case, because the failure of both the powertrain and the auxiliary battery is required to cause the specified hazard, the stringency of the ASIL allocated to each of these failures can be reduced according to the given algebra.

$$\sum_{j=1}^{i} ASIL_{component_j} \geq ASIL_{hazard} \qquad (1)$$

To facilitate this, each of the ASILs can be represented by an integer value 0 to 4 as shown in Table 3.

Table 4 shows all the combinations of ASILs that could be decomposed to the powertrain and auxiliary battery failures respectively along with the algebraic values for each of those ASILs. The final column shows the sum value of the two algebraic values. Where the sum value equals or exceeds 3 (the algebra value associated with ASIL C) the decomposition is deemed to be valid.

The last 6 combinations have a sum value of less than three so can be discarded as invalid decompositions. The four shaded rows show the combinations that exactly meet the requirement. The remaining 15 rows also exceed the stringency of the safety requirement. These can be considered a valid decomposition, however it is likely to be suboptimal once cost is considered as generally delivering a function at a higher safety integrity level is more costly.

If we consider this one hazard, then we can be satisfied that if any of the shaded combinations from the table are chosen, then we will meet the requirements of avoiding the hazard. However, the reality is more complicated.

This cut set is shared across multiple hazards. One of these is the "Loss of Braking All Wheels" hazard that was assigned ASIL D. When we include this constraint, the shaded combinations are no longer valid as their sum value is less than 4, the algebraic value for ASIL D.

There are 5 combinations that exactly meet the ASIL D requirement, but further factors need to be considered before making a final selection.

The cut set under consideration is of order 2 and contains the failure of the auxiliary battery and the powertrain battery. The auxiliary battery failure is part of an additional 9 order 3 cut sets of the "Loss of Braking Rear" hazard.

As an example we can consider one of these cut sets: omission failure of the auxiliary battery and the IWM of brake unit 3 and the IWM of brake unit 4. The decomposition that we choose for this cut set is affected by the choice of decomposition from the previous cut set. If we chose to allocate ASIL D to the auxiliary battery in the previous cut set, then we could potentially allocate QM to each of the omission failures of the IWMs of brake unit 3 and 4. However, if we had chosen one of the other decompositions such as QM to the auxiliary battery (and ASIL D to the powertrain battery), then the stringency of the decompositions to the other failures in the second cut set would have needed to be higher to meet the requirements.

If we also consider the "Loss of Braking All Wheels" hazard that adds another 81 order 5 cut sets. Then it is necessary to also consider the auxiliary battery's contribution to 3 other hazards and all of their cut sets. Similarly, the choice of decomposition to the first cut set pair also has knock on effects for any and all cut sets that contain the powertrain battery.

The ASIL algebra provides a way of determining the validity of a given decomposition. There are however additional factors that will influence the choice of ASIL combinations when decomposing in a system. A significant one is development cost. The ASIL allocated to a component represents the stringency of requirements that need to be complied with when developing it. Therefore, the higher the ASIL the higher the development cost. Where the safety requirements can be met, it is desirable to find an decomposition of ASILs that minimizes the cost of doing so.

It is often the case in the early stages of the design process, that the precise development costs of the components or functions in a system cannot be provided. That does not mean that cost cannot be considered as part of the decomposition process. In lieu of individual component costs, it is possible to consider the relative expected cost of development. In the simplest case, the algebra values in table 3 can be used as a linear cost, but this does not serve in further distinguishing the different combinations of decompositions. Table 5 provides a non-linear cost heuristic based on the experiential observation that the difference in cost between ASIL B to ASIL C is greater than the difference between the other ASILs [10].

Further exploration of the application of different cost heuristics to the optimization of ASIL decomposition can be found here [11].

When you apply this cost heuristic to the decomposition combinations available for the auxiliary battery and powertrain battery cut set in the "Loss of Braking Rear Wheels" hazard you get the results shown in Table 6. According to the heuristic, the shaded combinations are less costly than the unshaded combinations despite both meeting the safety requirement for that hazard.

It is important to remember that the failures in the cut sets are shared in multiple cut sets across multiple hazard fault trees. Therefore, in order to calculate the cost for the system it is summed once per failure and not once for every occurrence in the cut sets.

It is clear that achieving valid ASIL decompositions at minimal cost across a system manually is a practically impossible task. The many possible combinations, the multiple constraints provided by the hazards, and the knock-on effect of the interconnected fault trees and cut sets, leads to a combinatorial explosion. This is an optimization problem that requires the use of automated optimization algorithms to solve.

To do this it is necessary to encode potential solutions to the ASIL decomposition problem. The problem that is to be solved is: what is the ASIL requirement of each component/failure mode in the system such that the requirements assigned to the hazards are satisfied; and at minimum cost. The encoding that the algorithm can work with is a list

TABLE 5 This table shows the experiential cost heuristic for each ASIL

ASIL	Cost
QM	0
A	10
B	20
C	40
D	50

TABLE 6 This table shows the estimated cost of possible choices for decomposition due to the powertrain and auxiliary battery cut set for the "loss of braking rear wheels" hazard

Powertrain battery		Auxiliary battery		
ASIL	cost	ASIL	Cost	Total cost
A	10	B	20	30
B	20	A	10	30
C	40	QM	0	40
QM	0	C	40	40

FIGURE 2 Example solution encoding at iteration t showing the ASILs allocated to each unique failure mode in the system.

Solution^t	FM1	FM2	FM3	FM4	FM5
	B	D	C	B	A

of all the unique failure modes in the system and the ASIL that has been allocated to it. An example of this is shown in Figure 2.

An encoding can be validated against the hazards' ASILs by considering each cut set in turn, summing the ASIL algebra values of each of the failure modes from the cut set (as provided by the encoding list) and noting whether the sum result is equal to or exceeds the value of the current cut set's hazard ASIL. If this is true for all of the cut sets of all of the hazards, then the current encoding is valid.

The cost of a solution is calculated by looking up the cost (such as in Table 5) of each allocated ASIL in the encoding and summing them together to provide the total ASIL related costs of the system. The example shown in Figure 2 has the ASIL cost of 140 (20 + 50 + 40 + 20 + 10).

Tabu Search

The optimization technique applied for this paper uses a Tabu search variant algorithm [7]. It is based on the Steepest Ascent Mildest Descent (SAMD) method used by Hansen and Lih [12] for their work on system reliability optimization. One modification made for the ASIL decomposition problem is to adapt the method to a Steepest Descent Mildest Ascent (SDMA) as the algorithm seeks to minimize the development costs associated with the safety requirements, rather than the maximization objective of the SAMD approach.

The SDMA method attempts to follow the steepest descent path through the search space until a local minimum is detected. In order to escape the local minima, the algorithm uses the mildest ascent route available to it.

In order to achieve the steepest descent during an iteration of the algorithm it is necessary to choose a failure mode from the encoding and reduce its decomposed ASIL by one (i.e. from ASIL C to ASIL B). The reduction in the chosen failure mode's ASIL should result in the largest reduction in system cost. In the case of the example shown in Figure 2 the chosen failure mode would be FM3. The cost difference of reducing from ASIL C to ASIL B is 20, whereas the cost difference of all of the other available reductions is 10 (ASIL D to C, ASIL B to A, and ASIL A to QM as given by the cost heuristic in Table 5). The resultant encoding is shown in Figure 3.

FIGURE 3 Example solution encoding at iteration t + 1 where the steepest descent was followed by reducing the ASIL of FM3, shown in bold.

Solution^{t+1}	FM1	FM2	FM3	FM4	FM5
	B	D	**B**	B	A

To demonstrate selecting the mildest ascent we will assume that the solution at iteration t + 1 in Figure 3 is in a local minimum. This can occur if it is not possible to reduce any of the ASILs in the solution without invalidating one or more of the hazards' safety requirements. To produce the mildest ascent, it is necessary to choose one of the failure modes and increase the ASIL of its safety requirement by one such that it results in the smallest increase in cost. In the case of our example we would select FM5 resulting in an increase in cost of 10. The other choices either result in an increase in cost of 20 (FM1, FM3, and FM4 from ASIL B to ASIL C) or cannot be increase further (FM2 ASIL D). The resultant encoding is shown in Figure 4.

FIGURE 4 Example solution encoding at iteration t + 2 where the mildest ascent was followed by increasing the ASIL of FM5, shown in bold.

Solution^{t+2}	FM1	FM2	FM3	FM4	FM5
	B	D	B	B	**B**

An adaptive memory structure (the Tabu list) is used to prevent the algorithm from making reverse moves and falling back in to local minima. A variable f_i (where i refers to the failure mode that was just increased) stores how many iterations a reverse move will be forbidden for. After making an ascent move this variable is set to a number of iterations *p*. Conversely, following a descent move, the variable f'_i is set to a number of iterations *p'* and stores for how many iterations the failure mode will be blocked from increasing.

The use of such a memory structure increases the diversity in the search by forcing the algorithm to be more explorative. In order to decrease the algorithms sensitivity to the initial selection of the *p* and *p'* values, they are adjusted dynamically, incrementing at intervals $updatePeriod_p$ and $updatePeriod_{p'}$ respectively. When they reach their maximum values $limit_p$ and $limit_q$, they are reset to zero.

The algorithm includes an aspiration criterion which allows it to make a move forbidden by the memory structure if the resultant solution will be superior to any found previously.

Figure 5 summarizes the SDMA Tabu search algorithm used in this paper.

FIGURE 5 Tabu search overview.

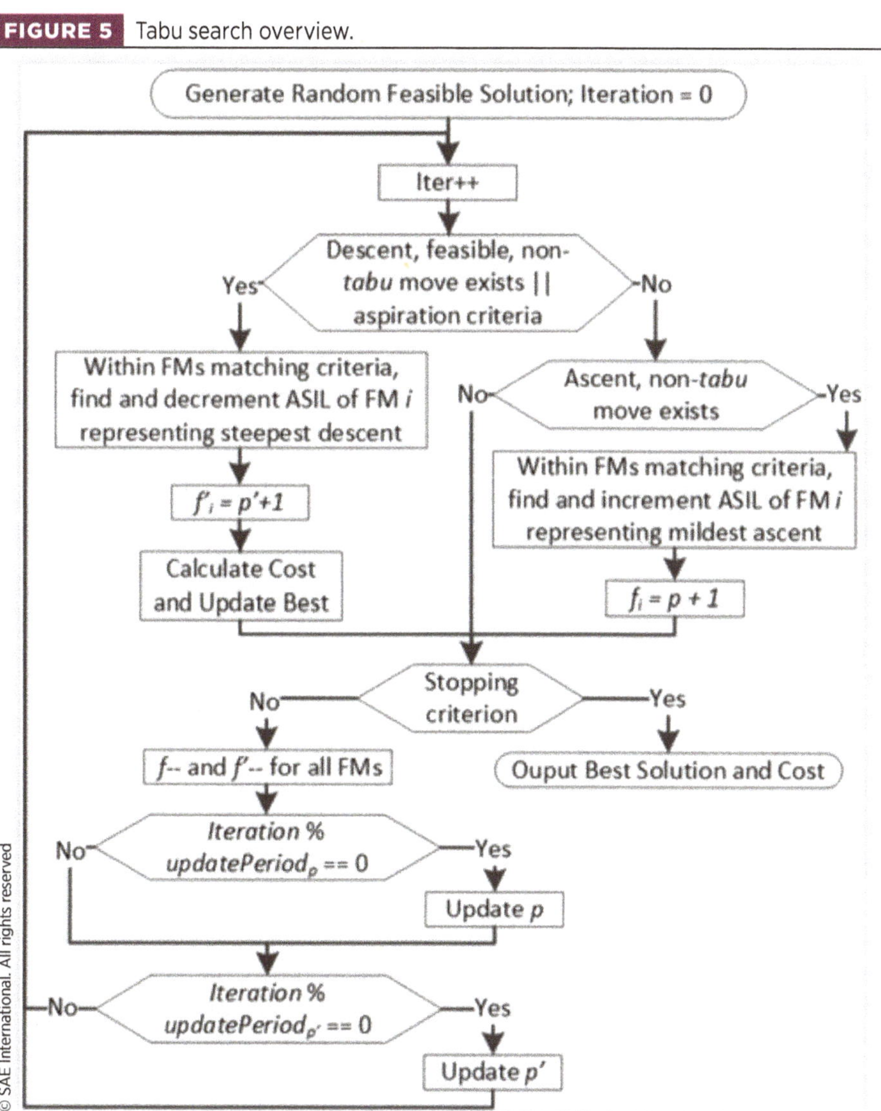

Failure Modes Versus Components

Earlier work with the HiP-HOPS ASIL decomposition techniques used the Tabu search algorithm as described in this paper. The encoding for the search algorithm stores an ASIL value for each of the failure modes in the system. It is theorized that taking advantage of the automatic fault tree analysis at the granularity of the failure modes allows for the specification of safety requirements for the development of (sub-) systems and their components that would be superior (less costly) than if forced to allocate at the component level. This approach considers ASIL decomposition at a level that is not described in the ISO 26262 standard, which speaks only of decomposing down to the level of component.

This paper takes a closer look at the consequences of such a limit in terms of the solutions possible when decomposing down to the failure modes as compared to different approaches for achieving this at the component level.

The first approach being considered is a naïve conversion. This involves running the previous ASIL decomposition algorithm to allocate ASILs to the failure modes of the system. The failure modes can be traced back to the system model that generated them. This means that for each component in the system, it is possible to collect the highest ASIL that was decomposed to one of its failures.

For example, in the HBS case study, the auxiliary battery component has two failure modes: an omission and a value failure. As a result of the optimization algorithm, they are allocated ASIL B and ASIL D respectively. Selecting the highest of these values results in us allocating ASIL D to the auxiliary battery component.

The second approach involves altering the optimization algorithm so that the encoding of the solution is not a list of ASILs decomposed to each of the failure modes in the system, but rather at the less granular level of the components. The algorithm manipulates the allocated ASILs in the encoding in the same manner as before. However, in order to establish the validity of the decomposition it is necessary to associate the components ASIL with all of its failure modes. These in turn are then used to validate the decompositions through the cut sets as before.

An example of this would be that if the ASIL allocated to the auxiliary battery in the solution encoding was ASIL C, then both the omission failure and the value failure of that component would be set to ASIL C. The validity check would reveal this to be an invalid decomposition due to one or more of the hazards' ASIL requirement.

To calculate the development cost in both approaches to optimizing the ASIL decomposition at the component level, the value is calculated by summing the heuristic cost of the effective component faults' ASILs. For example, the cost of setting the auxiliary battery ASIL to C is 80 because it has two failure modes that both derive their ASILs from the parent component. It is done this way for this paper so that the resultant cost can be directly compared from all three approaches.

Results

Table 7 shows the results of running the optimization algorithm in each of the three approaches. The first column indicates the 60 unique failure modes in the Hybrid braking case study. The naming convention used here gives first the name of the component followed by the name of the failure mode separated by a period. For example, EMB1. Omission refers to the omission failure of the electro-mechanical brake in the first wheel brake module.

TABLE 7 This table shows the decomposed ASILs for the failure modes of the system when using the different decomposition techniques. The FM column shows the original HiP-HOPS technique that decomposes to the failure modes in the system. The FM- > C column post-processes the ASILs to assign the highest sub-value to each component. The C column optimizes directly to the components. The cells marked in grey highlight differences between the latter two results

Failure Mode	Hazard	ASILs allocated per:		
		FM	FM- > C	C
Battery_Aux.Omission	2-5	B	D	D
Battery_Aux.Value	1	D	D	D
Battery_PT.Omission	2-5	B	B	B
Battery_PT.Value	1	B	B	B
Communication_Bus1.Omission	2-5	B	D	D
Communication_Bus2.Omission	2-5	B	D	D
Electronic_Pedal.Omission1	1-6	D	D	D
Electronic_Pedal.Omission2	2-5	B	D	D
Electronic_Pedal.Value1	1	D	D	D
Electronic_Pedal.Value2	1	D	D	D
Mechanical_Pedal.Omission	1-6	B	D	D
Mechanical_Pedal.Value	1	D	D	D
EMB1.Omission	3-6	QM	QM	B
EMB1.Value	1	QM	QM	B
EMB1_Power_Converter.Omission	3-6	QM	QM	B
EMB1_Power_Converter.Value	1	QM	QM	B
IWM1.Omission	3-6	A	A	B
IWM1.Value	1	QM	A	B
IWM1_Power_Converter.Omission	3-6	QM	A	B
IWM1_Power_Converter.Value	1	A	A	B
WNC1.Omission1	3-6	QM	A	B
WNC1.Omission2	3-6	A	A	B
WNC1.Value1	1	QM	A	B
WNC1.Value2	1	QM	A	B
EMB2.Omission	3-6	B	B	QM
EMB2.Value	1	B	B	QM
EMB2_Power_Converter.Omission	3-6	B	B	QM
EMB2_Power_Converter.Value	1	B	B	QM
IWM2.Omission	3-6	A	B	QM
IWM2.Value	1	B	B	QM
IWM2_Power_Converter.Omission	3-6	B	B	QM
IWM2_Power_Converter.Value	1	A	B	QM
WNC2.Omission1	3-6	B	B	QM
WNC2.Omission2	3-6	A	B	QM
WNC2.Value1	1	B	B	QM
WNC2.Value2	1	B	B	QM
EMB3.Omission	2,4-6	A	A	B
EMB3.Value	1	A	A	B
EMB3_Power_Converter.Omission	2,4-6	A	A	B
EMB3_Power_Converter.Value	1	A	A	B

TABLE 7 (*Continued*) This table shows the decomposed ASILs for the failure modes of the system when using the different decomposition techniques. The FM column shows the original HiP-HOPS technique that decomposes to the failure modes in the system. The FM- > C column post-processes the ASILs to assign the highest sub-value to each component. The C column optimizes directly to the components. The cells marked in grey highlight differences between the latter two results

		ASILs allocated per:		
Failure Mode	**Hazard**	**FM**	**FM- > C**	**C**
IWM3.Omission	2,4-6	QM	A	B
IWM3.Value	1	A	A	B
IWM3_Power_Converter.Omission	2,4-6	A	A	B
IWM3_Power_Converter.Value	1	QM	A	B
WNC3.Omission1	2,4-6	A	A	B
WNC3.Omission2	2,4-6	QM	A	B
WNC3.Value1	1	A	A	B
WNC3.Value2	1	A	A	B
EMB4.Omission	2,4-6	A	A	QM
EMB4.Value	1	A	A	QM
EMB4_Power_Converter.Omission	2,4-6	A	A	QM
EMB4_Power_Converter.Value	1	A	A	QM
IWM4.Omission	2,4-6	A	A	QM
IWM4.Value	1	A	A	QM
IWM4_Power_Converter.Omission	2,4-6	A	A	QM
IWM4_Power_Converter.Value	1	A	A	QM
WNC4.Omission1	2,4-6	A	A	QM
WNC4.Omission2	2,4-6	A	A	QM
WNC4.Value1	1	A	A	QM
WNC4.Value2	1	A	A	QM
Total cost		840	1100	1020

The second (Hazard) column shows the hazards (indexed in table 1 with its ASIL) that the failure mode contributes to through the (many) cut sets (shown in table 2). In all cases the failure mode contributes (at least indirectly) to a hazard with ASIL D. The third (FM) column shows the ASIL that is allocated using the pure direct to failure mode optimization approach. The fourth (FM- > C) column shows the ASIL that are derived from assigning the highest ASIL from the first approach to the parent component of each of the failure modes. The final column (C) shows the ASIL that is allocated when the optimization algorithm decomposes the ASILs directly to the components of the system.

At the bottom of the table the ASIL development cost is noted for each of the three approaches.

The shaded cells in the last column highlight where the allocations made by the two different component focused algorithms are different.

What these results show is that the finer granularity of allocating down to the level of the failure modes allows the algorithm to find a more cost-effective solution. This would seem to be highly desirable in situations where vendors would be able to deliver components that can meet specific safety requirements for the different failure modes of the parts, or where the component is effectively a subsystem and adequate partitioning can be established between elements within.

Where this is not possible it is necessary to specify the safety requirements at the level of the components, which is more in keeping with the process as laid out by the ISO 26262 standard. Here, using the failure mode allocation technique and converting the results to the component level produces and inferior, less cost-effective solution that optimizing directly to the components using the specialized algorithm.

In the latter approach the components of each wheel brake module are treated more uniformly and because they represent independent redundancy the distribution of the ASILs is more favorable.

This is not the end of the story however as the ability of the direct to component allocation algorithm to find superior solutions to the conversion approach depends on the cost heuristic being used. If, for example, a logarithmic cost heuristic is used like that in Table 8, then the solution identified by the two component focused approaches is the same. This is shown in Table 9.

TABLE 8 This table shows an alternative logarithmic cost heuristic for each ASIL

ASIL	Cost
QM	0
A	10
B	100
C	1000
D	10000

TABLE 9 This table shows the costs of running the different optimization approaches with a logarithmic cost heuristic such as in Table 8

ASILs allocated per:		
FM	FM - > C	C
51150	100680	100680

In order for the direct to component optimization to find superior solutions, it is necessary for the cost heuristic to have moves between different ASILs to have interchangeable cost differences. For example, with the cost heuristic shown in Table 5 only the jump from ASIL B to ASIL C is unique (20 units compared to 10 for all the other jumps). The logarithmic cost heuristic in Table 8 has unique cost jumps for all of its ASILs. It should be noted that the direct to failure mode approach finds markedly superior solutions in all cases.

An additional consideration is the performance cost. When optimizing directly to the components the search space is considerably reduced. There are 60 failure modes in the case study system but only 24 components. The direct to components algorithm took a little over a second to complete one run of the algorithm compared to just under 9 seconds to run the direct to failure modes algorithm.

With these different factors in consideration it appears that the obvious choice when constrained to consider ASILs at a component level only is to use an algorithm that specially targets that objective directly. It is quicker, and the resultant configuration of ASIL allocations may be superior.

However, if it is possible to consider the allocation of ASILs to the more granular level of the failure modes of a system, then a more cost effective solution is likely to be found.

Conclusions

The safety engineering approach described in the automotive standard ISO 26262 requires the consideration of safety right from the early stages of the design process. One of the key pillars of this are the ASILs that can be assigned to the safety requirements of the system. Importantly, these requirements can then be distributed throughout the components of the system and decomposed where independent redundancy can be shown to manage the cost of meeting these requirements.

There is additional effort/cost required due to decomposition (for example, proof of independence needed) which is not considered in this study. This cost likely not negligible and it would be worth estimating these costs in the future. However, decomposition is precisely used in order to reduce costs so the relative cost of decomposition in general must be significantly lower than the benefits of reducing ASILs.

Doing this manually, even in small systems is impractical to the point of being impossible if the expectation is to achieve cost optimality. Automated systems are necessary to cover the vast search spaces that are generated by the combinatorial explosion of potential configuration.

This paper described the recent work in this area implemented in the HiP-HOPS safety analysis and optimization tool. Two modes of operation are shown, allocation to components as intended by the ISO 26262 standard, and the theoretical allocation down to the level of component failure modes.

The approach described here is not a 'fire and forget', one-time application to provide automatic safety standard compliance. Rather it should be considered as an assistive technique to help inform engineer choices in their efforts for cost-effective standard compliance; one that can be applied iteratively throughout the design life of a system.

Comparison of the two modes reveals the economic benefits available where we are able to use the latter, more granulated allocation process. Where this is not possible specialized component focused algorithms offer potential advantages over simply converting the results. In all cases, it is more efficient working with a smaller search space, and in some cases may provide superior, more cost effective solutions, though this will depend on the cost heuristic being used.

Contact Information

David Parker
d.j.parker@hull.ac.uk
University of Hull, UK

References

1. Int'l Organization for Standardization, "ISO 26262 Road Vehicles—Functional Safety," 2011.
2. Ward, D.D. and Crozier, S.E., " The Uses and Abuses of ASIL Decomposition in ISO 26262. in System Safety, Incorporating the Cyber Security Conference 2012," *7th IET International Conference on,* 2012.
3. Papadopoulos, Y., Walker, M., Reiser, M-O., Servat, D. et al., "Automatic Allocation of Safety Integrity Levels, *8th European Dependable Computing Conference—CARS Workshop*, Valencia, Spain, 2010, ACM Press, 7-11, ISBN:978-1-60558-915-2.
4. Mader, R., Armengaud, E., Leitner, A., and Steger, C.. "Automatic and Optimal Allocation of Safety Integrity Levels," *Reliability and Maintainability Symposium (RAMS 2012)*, Reno, NV, 2012, 1-6, doi:10.1109/RAMS.2012.6175431.
5. Murashkin, A., Silva Azevedo, L., Guo, J., Zulkoski, E. et al., "Automated Decomposition and Allocation of Automotive Safety Integrity Levels Using Exact Solvers," *SAE Int. J. Passeng. Cars - Electron. Electr. Syst.* 8, no. 1 (2015): 70-78, doi:10.4271/2015-01-0156.
6. Parker, D., Walker, M., Azevedo, L., Papadopoulos, Y. et al., "Automatic Decomposition and Allocation of Safety Integrity Levels Using a Penalty-Based Genetic Algorithm," *Proceedings of the 26th International Conference on Industrial, Engineering, and Other*

Applications of Applied Intelligent Systems (IEA/AIE 2012): Special Session on Decision Support for Safety-Related Systems, Amsterdam, the Netherlands, June 17-21, 2013.

7. Silva Azevedo, L., Parker, D., Walker, M., Papadopoulos, Y. et al., "Automatic Decomposition of Safety Integrity Levels: Optimisation by Tabu Search," *2nd Workshop on Critical Automotive Applications: Robustness & Safety (CARS), at the 32nd International Conference on Computer Safety, Reliability, and Security (SAFECOMP'13)*, Toulouse, France, 2013.
8. Azevedo, L.P., "Hybrid Braking System for Electrical Vehicles: Functional Safety," M.Sc. thesis, Dept. Elect. Eng., Porto Univ., Porto, Portugal, 2012.
9. de Castro, R., Araújo, R.E., and Freitas, D., "Hybrid ABS with Electric Motor and Friction Brakes," presented at *the IAVSD2011 - 22nd International Symposium on Dynamics of Vehicles on Roads and Tracks*, Manchester, UK, 2011.
10. Allen, M., "Cost versus ASIL," February 2, 2012, ISO 26262 Functional Safety [LinkedIn], accessed May 1, 2014, http://www.linkedin.com/groups/Cost-versus-ASIL-2308567.S.92692199?view=&srchtype=discussedNews&gid=2308567&item=92692199&type=member&trk=eml-anet_dig-b_pd-ttl-cn&ut=1evtvoEm1QcBw1.
11. Azevedo L.S., Parker D., Papadopoulos Y., Walker M. et al., Exploring the Impact of Different Cost Heuristics in the Allocation of Safety Integrity Levels, Ortmeier F. and Rauzy A. (eds.), *Model-Based Safety and Assessment. Lecture Notes in Computer Science*, Vol. 8822 (Cham: Springer, 2014).
12. Hansen, P. and Lih, K.-W., "Heuristic Reliability Optimization by Tabu Search," *Annals of Operations Research* 63 (1996): 321-336.

CHAPTER 7

An Integrated Approach to Requirements Development and Hazard Analysis

John Thomas, John Sgueglia, Dajiang Suo, and Nancy Leveson
Massachusetts Institute of Technology

Mark Vernacchia and Padma Sundaram
General Motors Company

The introduction of new safety critical features using software-intensive systems presents a growing challenge to hazard analysis and requirements development. These systems are rich in feature content and can interact with other vehicle systems in complex ways, making the early development of proper requirements critical. Catching potential problems as early as possible is essential because the cost increases exponentially the longer problems remain undetected. However, in practice these problems are often subtle and can remain undetected until integration, testing, production, or even later, when the cost of fixing them is the highest.

In this paper, a new technique is demonstrated to perform a hazard analysis in parallel with system and requirements development. The proposed model-based technique begins during early development when design uncertainty is highest and is refined iteratively as development progresses to drive the requirements and necessary design features. The technique is evaluated by applying it to a realistic but generic Shift-By-Wire design concept in two iterations with varying levels of detail. In addition, as the requirements and design evolve and change over time, the changes can be immediately analyzed for new hazards without repeating the entire analysis. The approach is also applicable even before requirements are developed, providing feedback when some of the most important decisions are being made instead of waiting for a finished design or model to begin an analysis. In this way, potential issues can be identified immediately and more efficiently, thereby reducing the need for future rework.

CITATION: Thomas, J., Sgueglia, J., Suo, D., Leveson, N. et al., "An Integrated Approach to Requirements Development and Hazard Analysis," SAE Technical Paper 2015-01-0274, 2015, doi:10.4271/2015-01-0274.

1. Introduction

Modern automobiles are incorporating more and more advanced software-controlled safety features such as lane keeping assist, automatic emergency braking, adaptive cruise control, and a growing number of automated or semi-automated systems. Increasingly complex behaviors are now being incorporated into safety-critical software and there are very few physical constraints that limit the complexity of the software elements. While this has allowed many innovative features that were not possible in the past, the additional complexity has made it much more challenging to develop these systems and perform adequate safety analyses. Software-related issues are frequently being discovered late in development, testing, and even during production when the problems are the most expensive to fix and when the range of practical solutions is the most limited.

The software in modern vehicles is not only complex but also increasing in size exponentially, making the need for efficient analysis techniques increasingly urgent. One manufacturer reports two to three million lines of software code for average 2005 models [1], six million lines of code for 2007-2008 models [1], and sixteen million lines of code in 2013 [2]. Meanwhile, NHTSA data shows that more and more software-related issues are being discovered too late. About 382,000 U.S. vehicles were indicated in software-related recalls in the year 2000, compared to 13 million in 2013 and 48 million a year later [3]. In fact, some have claimed that up to 50% of car warranty costs are related to the electronics and their embedded software [4].

The traditional techniques used for analysis have not kept pace with the increased complexity of modern software-intensive systems. Although most analysis techniques consider individual failures (e.g. Fault Tree Analysis, Failure Modes and Effects Analysis, etc.), many other safety-critical decisions and assumptions are made during the development process and can easily be overlooked in a component failure-based analysis. Judgments about what behavioral requirements are needed, what interactions could potentially be dangerous, what information and feedback the software decisions should be based on, how to coordinate multiple distributed controllers, and other issues must be determined during development. While these issues are typically addressed using engineering best-judgment, they are often resolved in a compartmentalized manner without a complete understanding of the system-level impact of the proposed solution. As a result, unintended interactions are often introduced and may be overlooked by standard failure-based analyses.

In fact, a new class of accidents called "component interaction accidents" is becoming increasingly common. Component interaction accidents arise due to dysfunctional or unintended interactions among several components and often occur without a component failure. For example, the software requirements may be incorrect, incomplete, or ambiguous. Individual software components may operate exactly as required and as designed without failure, but the overall behavior of multiple interacting components may still lead to unanticipated vehicle behavior. In fact, most software-related accidents are caused not by coding errors but flawed software requirements [5, 6, 7, 12]. Such flaws can be very difficult to identify or anticipate with traditional failure-based methods or by analyzing individual components in isolation.

Another limitation is that most approaches to safety analysis focus on assessment of an existing design as opposed to driving the design and requirements from the start. Most techniques require a considerable amount of information about the system to perform an effective analysis, but some of the most important decisions affecting safety are made early during development before the design and requirements are known. When a safety analysis is eventually performed, the result is often rework or inefficient patches to correct earlier mistakes. In many cases the best solutions are no longer practical to implement by the time problems are discovered, resulting in difficult compromises.

Although system engineering processes continue to emphasize more up front hazard and risk assessments, integrating hazard analysis techniques into early engineering design processes remains a challenge. Engineering and analysis often remain separate sequential activities, effectively adding new analysis and assessment tasks to an already long list of otherwise unchanged engineering tasks. What is needed is not simply more analysis and more paperwork, but more efficient and effective ways to do system engineering from the start. A truly integrated safety-driven design process should not just periodically monitor the design but continuously drive decisions as they are made so that safety is "baked in" and less analysis is required later. Such a process has the potential to identify problems sooner, reduce rework when problems are corrected, reduce the total integration time needed later, and simplify the final safety assessment by leveraging analysis artifacts already created during development.

The functional safety standard ISO 26262 describes activities and requirements throughout the safety lifecycle of safety-related systems comprised of electrical, electronic, and software elements that provide safety-related functions [8]. The standard is based on the popular system engineering V-Model [9] and has the potential to encourage consideration of safety aspects much earlier than other standards such as IEC 61508 [10, 11]. In many areas, ISO 26262 lists broad activities that must be done but does not prescribe a specific method to be used or provide a specific process to follow. For example, Part 3 Clause 7.4.4.2.1 [8] requires that potential sources of harm "shall be determined systematically by using adequate techniques". A note suggests that a variety of techniques such as brainstorming could be used, but no specific process is provided to ensure the requirement is met. Part 3 Clause 7.4.4.2.4 [8] requires that "consequences of hazardous events shall be identified.", but no method or process is indicated and "relevant" is left to the reader to decide. These are not necessarily weaknesses of the standard, and in fact these are advantages in many ways because they offer flexibility and can accommodate innovation and advancement as new hazard analysis methods are developed.* However, the standard is not enough on its own as it generally stops short of prescribing a specific method or technique to be used.†

Hazard analysis methods such as STPA (System Theoretic Process Analysis) can be used to satisfy some of these goals. STPA was created based on systems theory to address the limitations of traditional safety analysis techniques [12]. STPA is a top-down hazard analysis method designed to go beyond traditional component failures to also identify problems such as dysfunctional interactions, flawed requirements, design errors, external disturbances, human error and human-computer interaction issues, and other problems. Although STPA has been very successful, it is typically applied as a separate analysis-i.e. all steps of the traditional STPA analysis are completed before the system is fixed, refined, or augmented. More detailed comparisons and evaluations of STPA can be found in [13, 14, 15].

---

* In other areas ISO26262 provides a non-exhaustive list of methods that might be used, but stops short of requiring a specific method. For example, Part 9 Clause 8.2 [8] lists Failure Modes and Effects Analysis (FMEA), Fault Tree Analysis (FTA), HAZOP, and Event Tree Analysis (ETA), although with a note that "The qualitative analysis methods listed above can be applied to software where no more appropriate software-specific analysis methods exist." Part 4 Clause 7.4.3.1 [8] provides recommendations for inductive vs. deductive analysis methods, but does not require a particular method to be used.

† For a more detailed assessment of ISO 26262 and its relationship to system engineering and System Theoretic Process Analysis see [11].

This paper proposes and demonstrates a new safety-guided *design* methodology based on STPA that interleaves development and analysis tasks to provide an integrated development process. The process begins with very little information about the system design and proceeds to drive an initial control model and initial behavioral requirements. Design decisions can be incorporated in top-down fashion as they are made, and the safety implications are immediately fed back into the engineering process to reduce future rework. New techniques are also incorporated to help systematically ensure potential unsafe controls are not overlooked and that the resulting behavioral requirements are complete and consistent. This process can also be used to help prevent initial mistakes in requirements, specification, and design rather than waiting to assess and correct mistakes after-the-fact. In addition, all intermediate results of the process can be accumulated and assembled at the end to create a comprehensive overall safety analysis, thereby reducing extra work that is often done at the conclusion of each development stage.

2. System Theoretic Process Analysis

STPA begins by identifying the system accidents and system hazards to be prevented [12]. In STPA, an accident is an event that results in a loss and a system hazard is a system state that will lead to an accident in a worst-case environment [12]. Although the accidents of interest often involve human injury or loss of life, STPA can also be applied more generally to other losses that must be prevented [12]. Other losses may include quality issues such as loss of customer satisfaction, performance issues such as reduced power or efficiency, economic issues such as damage to the vehicle, environmental pollution, or any other loss that is unacceptable.

STPA uses a model of the control relationships within the system to guide the analysis. These relationships are modeled using a safety control structure, as shown in Figure 1. Control actions or commands that affect lower-level processes are identified as well as feedback that is used to inform higher-level controllers. The analysis begins by identifying control actions that can be unsafe and may lead to a hazard. There are four ways in which a control action may be hazardous:

1. A control action required for safety is not provided.
2. A control action is provided when it is unsafe to do so.
3. A control action is provided too early or too late.
4. A continuous control action is applied too long or stopped too soon.

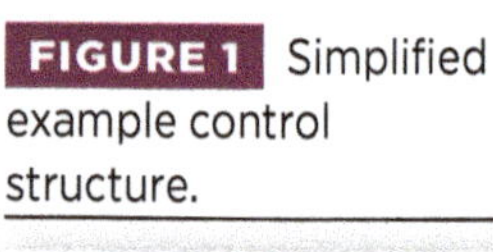

FIGURE 1 Simplified example control structure.

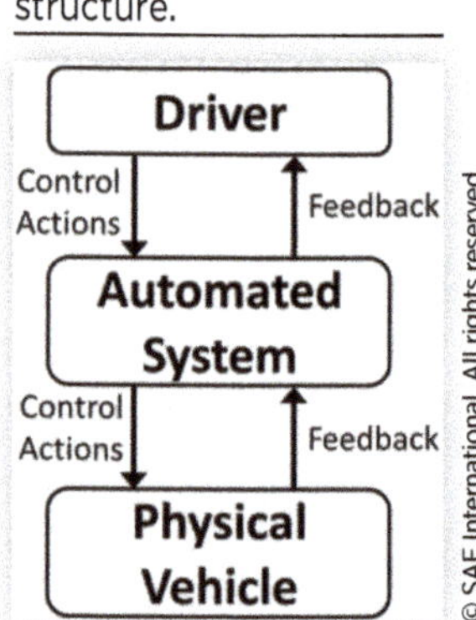

Once Unsafe Control Actions (UCAs) have been identified, the analysis proceeds to identify potential causes of the unsafe control. A generic control loop with causal factors, as shown in Figure 2 from [12], can be used to guide this part of the analysis. One important cause of unsafe control actions involves the controller's process model, also called a mental model when the controller is a human. The process model captures the controller's understanding and beliefs about the outside world, including the state of the controlled process and assumptions about how the controlled process works. Accidents in complex systems, particularly those related to software, often result from inconsistencies between the model of the process used by the controller and the actual process state, which leads to the controller providing unsafe control. Figure 2 also shows other potential causes such as missing or inadequate feedback, component failures (such as a sensor failure), inadequate control inputs, and others.

In addition to identifying causes of unsafe control actions, STPA also identifies how safe control actions may not be followed or executed properly. For example, appropriate

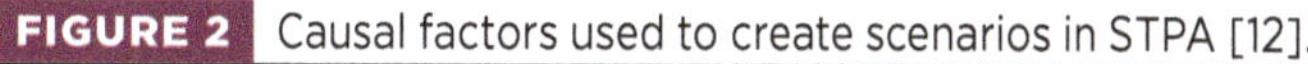

FIGURE 2 Causal factors used to create scenarios in STPA [12].

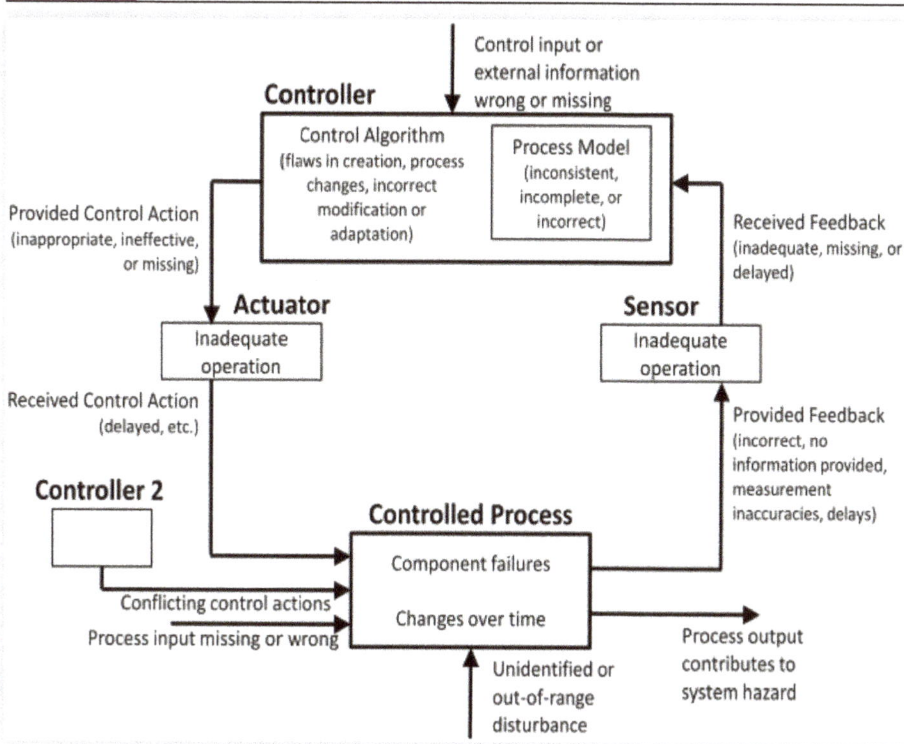

control actions may be issued but the actuator may introduce excessive delays or the controlled process may experience a physical failure that prevents the actuator from being effective.

The following sections introduce a safety-driven design process based on STPA that begins with very little information and proceeds to drive initial development and design decisions. The process is demonstrated using a realistic but fictional shift-by-wire concept with the potential for safety-critical automated behaviors.

3. New Integrated Approach

3.1 Process Overview

The safety-driven design process proposed in this paper interleaves hazard analysis tasks with development tasks to provide quicker feedback to engineers and to provide guidance as the system is being developed. The approach can be summarized as:

First iteration:

1. Define the system accidents and system hazards
2. Create the initial control structure
3. Identify initial unsafe control actions
4. Derive safety constraints/requirements from unsafe control actions
   a. Use the safety constraints to revise the control structure and design
5. Identify high-level causal scenarios
   a. Identify controls to eliminate or mitigate the high-level scenarios

Second iteration:

6. Formalize the unsafe control actions to identify any missing or conflicting UCAs and constraints
   a. Resolve the identified conflicts and revise the safety constraints
7. For scenarios not already controlled, identify more detailed causes by incorporating additional design detail
   a. Provide controls for the new causal factors identified

3.2 System Overview

A new vehicle shift-by-wire concept was chosen to demonstrate and evaluate the proposed safety-driven design process. The generic shift-by-wire concept was chosen because it introduces a new software controller with the potential for automated behavior and new safety features. It is also a relatively new technology that is under active development by several manufacturers. In addition, there are a large number of design decisions that clearly impact safety (such as what automated behaviors are appropriate and what information the controller should monitor) but there are very few regulations, industry standards, or processes that provide clear answers.

The shift-by-wire concept replaces traditional mechanical cables between the shifter and the transmission with an electronic lever, a shift control module, and various electronic actuators and sensors. The shift control module senses the shift lever position and commands some actuator to achieve the appropriate transmission range. Although the shift-by-wire concept may reduce manufacturing costs and increase packaging flexibility, there are several important safety implications that require careful consideration. Note that many details-such as the shift control module algorithm and the method of sensing and actuating-are unknown at this early stage and they are not needed to begin the proposed safety driven design process.

3.3 Building a Foundation

The first step is to identify the system accidents and system hazards. These form a foundation for the safety-driven design process and define the losses or system states are unacceptable and must be prevented. At this early phase, the system is the overall vehicle. Tables 1 and 2 list the system accidents and system hazards defined for this case study. Notice that these are defined at a very high level and in fact they will be similar for many automotive systems. The system accidents and system hazards help define the scope of the effort, and all results from the safety-driven design process will be traceable to these system accidents and system hazards.

Note that the term "system hazard" as defined in STPA is different from the term "hazard" in ISO 26262, which is defined as any potential source of physical injury or damage to the health of people [12]. Because STPA is a top-down process, it does not start with a list of all potential sources of injury or damage. Instead, it begins with system-level states or conditions that must be prevented (i.e. system hazards) and proceeds to systematically identify the potential causes.*

TABLE 1 System accidents

Number	System Accident Description
A-1	Two or more vehicles collide
A-2	Vehicle collides with other obstacle or terrain
A-3	Vehicle occupants injured without vehicle collision

TABLE 2 System hazards

Number	System Hazard Description
H-1	Vehicle does not maintain safe distance from nearby vehicles
H-2	Vehicle does not maintain safe distance from terrain and other obstacles
H-3	Vehicle enters uncontrollable/ unrecoverable state
H-4	Vehicle occupants exposed to harmful effects and/or health hazards

* In fact, STPA system accidents and system hazards can be defined much broader than ISO 26262 and may include anything that is unacceptable to the user and must be prevented. For example, loss of customer satisfaction or damage to the vehicle (without human injury) could be included if desired [12].

The first system hazard describes vehicles that become too close to each other. Because the system hazards are system states and conditions by definition, they do not specify potential causes and the analysis is not restricted to certain types of causes. The system hazards simply specify the system behavior that must be prevented, and all relevant causes will be systematically identified later. Potential causes of H-1 could involve a vehicle accelerating into another vehicle, an unsecured vehicle rolling backwards towards another vehicle, a vehicle moving in an unintended direction, physical malfunctions, driver errors, confusing automation, etc.

System hazard H-2 is similar to H-1 but refers to vehicles that are too close to non-vehicular objects. Examples of other objects include pedestrians, animals, bikers, and guardrails. H-3 captures situations in which the driver may be unable to use the vehicle's systems to gain control. This includes cases in which the vehicle may be travelling too fast, the driver's commands are ignored, or a system such as braking or power steering is ineffective. H-4 captures additional problems that may occur even without nearby objects such as a vehicle rollover, excessive deceleration or acceleration, or a vehicle fire.

Note that the System Hazards can lead to the System Accidents. For example, H-1 can cause A-1, H-2 can cause A-2, H-3 can lead to any of the accidents, and H-4 can lead to A-3.

Figure 3 shows the high-level system control structure model of the traditional mechanical shifter system, while Figure 4 shows the high-level control structure model for the proposed shift-by-wire system. These control structures are defined directly from the initial shift-by-wire description above and do not yet contain much detail. This is intentional; by starting at a very high level, the process can begin even before the design and requirements are available and provide actionable results before rework is needed.

FIGURE 3 High-level control structure for traditional mechanical shifting.

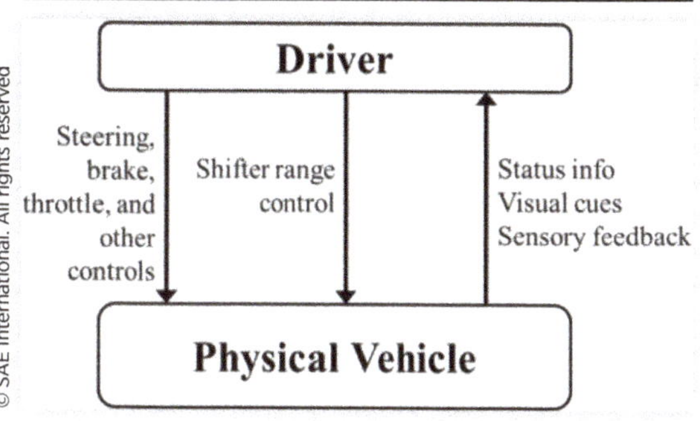

3.4 Initial Unsafe Control Actions and Safety Constraints

Once the high-level control structure is defined, an initial set of unsafe control actions can be identified by considering the four ways a control action can be unsafe, as discussed in the introduction. Table 3 shows the results that can be obtained at this stage of the process for the Range Command. The Range Command, also shown in Figure 4, is a command from the Shift Control Module to select a new range (e.g. Park, Reverse, Neutral, Drive, etc.).

Each unsafe control action is then translated into a safety constraint / safety requirement to be enforced, as shown in Table 4. As the design progresses, these unsafe control actions and safety constraints will be refined and examined in more detail.

Inconsistent means the requested range would cause physical damage, an unsafe change in motion, or violate motor vehicle regulations. Unavailable means a physical fault exists that would prevent the vehicle from shifting to the selected range. Also note that each UCA includes a link

FIGURE 4 Initial high-level control structure for shift-by-wire system.

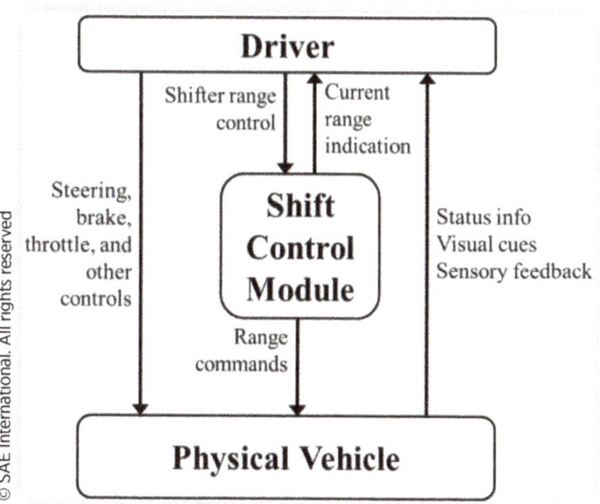

TABLE 3 Initial unsafe control actions for shifter control module

Control Action	Not providing causes hazard	Providing causes hazard	Too early, too late, wrong order	Stopped too soon, applied too long
Range command (from shifter control module)	UCA-1: Shifter Control Module does not provide range command when driver selects new range [H-1, H-2, H-3, H-4] UCA-2: Shifter Control Module does not provide new range command once current range becomes unavailable [H-1, H-2, H-3, H-4]	UCA-3: Shifter Control Module provides range command without driver new range selection [H-1, H-2, H-3, H-4] UCA-4: Shift Control Module provides range command when that range is unavailable [H- 3] UCA-5: Shift Control Module provides inconsistent range command [H-1, H-2, H-3, H-4]	UCA-6: Shifter Control Module provides range command too late after driver range selection [H-1, H-2, H-3, H-4] UCA-7: Shift Control Module provides range commands consistent with driver selection but in different order [H-1, H-2, H-3,H-4]	N/A

TABLE 4 Safety constraints derived from unsafe control actions

Safety Constraint	Description
SC-1	Shifter Control Module must provide range command when driver selects new range [H-1, H-2, H-3, H-4]
SC-2	Shifter Control Module must provide new range command once current range becomes unavailable [H-1, H-2, H-3, H-4]
SC-3	Shifter Control Module must not provide range command without driver new range selection [H-1, H-2, H-3, H-4]
SC-4	Shift Control Module must not provide range command when that range is unavailable [H-3]
SC-5	Shift Control Module must not provide range commands that are inconsistent [H-1, H-2, H-3, H-4]
SC-6	Shifter Control Module must provide range command within X milliseconds of driver range selection [H-1, H-2, H-3, H-4]
SC-7	Shift Control Module must provide range commands in the same order as received by the driver [H-1, H-2, H-3,H-4]

to the system hazards to provide traceability; every STPA result is traceable to the system hazards and accidents defined at the start of the analysis

As soon as the initial safety constraints are defined, potential design implications can be assessed. Although the process is not finished and causal factors have not yet been identified, these preliminary results can immediately be used to drive important design decisions and identify necessary safety features. The following questions can be used to guide this part of the process:

- Does the initial control structure allow the controller to monitor the conditions in the constraints?
- Do additional control actions need to be added to achieve or enforce the constraints?
- Are there other controllers that may interfere with or violate the constraints?

For example, SC-4 states that the shift control module must not issue a range command when that range is unavailable. How would the controller know which ranges are available or unavailable? The initial control structure in Figure 4 does not provide the Shift Control Module with that information. This new feedback can be immediately identified and added to the control structure without requiring additional design details or a detailed analysis. By identifying these dependencies early, solutions can be incorporated during early development without causing additional rework.

Figure 5 shows a revised control structure with the available range feedback. Note that it is not necessary yet to know how the available ranges are detected, just that they must be detected. The constraints produced at this stage are intentionally flexible to allow the design team to determine the most effective solutions. Once specific solutions are proposed, the control structure can be refined to include additional detail and identify whether any new safety concerns are introduced.

FIGURE 5 Revised control structure.

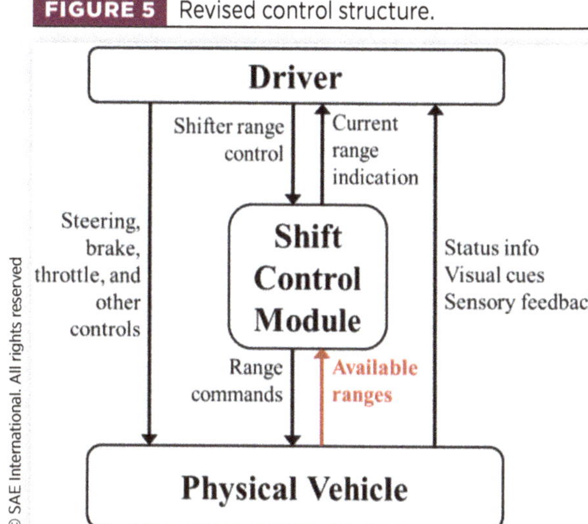

3.5 Identifying Causal Scenarios

Once the control structure has been revised, causal scenarios can be identified for each of the unsafe control actions. The causal factors in Figure 2 can be used to guide the generation of causal scenarios. Notice that more design information may be incorporated at this stage, such as information about the controller process model and other control inputs. Table 5 shows two causal scenarios that can be identified for the first unsafe control action and corresponding safety constraint:

Once the causal scenarios are identified, potential design implications can be assessed. The following questions can be used to help guide this part of the process:

- How does the controller determine the information referenced in the scenarios?
- Are additional controls needed to prevent identified flaws?
- Are new controllers or new functionalities needed?
- Do new constraints need to be defined?

For example, S-2 describes a case where the shift control module incorrectly believes a commanded range was achieved. How could this incorrect belief arise? In the control structure in Figure 5, the controller would only be aware of the last range command

TABLE 5 Causal scenarios that violate SC-1

Safety Constraint	Causal Scenario that Violates the Safety Constraint
SC-1: Shifter Control Module must provide range command when driver selects new range	**S-1:** Shifter Control Module does not provide range command because it *incorrectly believes no new range was selected*
	S-2: Shift Control Module does not send range command because it *incorrectly believes the range was already achieved*
	Etc.

FIGURE 6 Revised control structure.

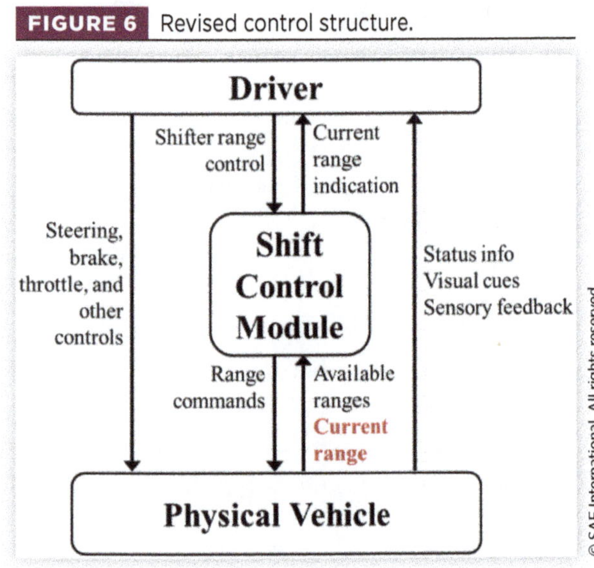

sent; there are no means for the controller to detect the current range. This reveals a potential hidden assumption that the current range matches the previous range command. However, if the command is not executed for any reason (whether a failure or otherwise), it will lead to S-2. One solution is to incorporate an additional feedback for the Shift Control Module to directly detect the current range, as shown in Figure 6.

By identifying these potential issues early, it is possible to drive the initial design decisions without causing rework.

So far, the process has focused on what control and feedback information will be required to be safe and what basic behaviors must be controlled. Specific design details have not yet been included so that the process can be performed quickly and extremely early during development. However, the design details eventually need to be incorporated into the process because they can introduce new safety concerns or cause unexpected side-effects. The next section explains how the process may be iterated to refine the requirements and provide more specific results based on the design details.

3.6 Refining and Formalizing Unsafe Control Actions

Although the initial set of unsafe control actions in Table 3 can be identified quickly to provide immediate feedback to the design team, a second iteration can be performed to help identify any additional UCAs and to identify and resolve any conflicts between the existing safety constraints. In addition, the results from the first iteration may have caused new interactions among existing controllers or new controllers to be introduced. By applying a formal framework to the initial set of unsafe control actions, a revised and more precise set of safety constraints and executable requirements can be defined.

A rigorous process for systematically identifying unsafe control using a formal framework has been developed in [16]. Missing UCAs and potential conflicts can be logically identified using a context table as shown in Table 6. Each row in a context table

TABLE 6 Context table for Shifter Control Module UCAs

Control Action	Driver Selected Range	SCM Selected Range Available	SCM Selected Range Consistent	Current range available	Not Providing Causes Hazards	Providing Causes Hazards	
Shifter Control Module provides range command when...	None	*	*	*	No	Yes	UCA-3
	*	*	*	No	Yes	No	UCA-2
	Doesn't match SCM cmd	*	*	*	No	Yes	
	Matches SCM cmd	*	*	*	Yes	No	UCA-1
	Matches SCM cmd	No	*	*	No	Yes	UCA-4
	Matches SCM cmd	*	No	*	No	Yes	UCA-5

* denotes conditions that do not matter for a given row

corresponds to an unsafe control action, and the columns represent different elements of the unsafe control action. For example, the first row of the context table in Table 6 corresponds to UCA-3 (Shifter Control Module provides range command without driver new range selection). The first column specifies that the control action is the range command. The next four columns describe the conditions that can make a range command unsafe, and asterisks are used to denote which conditions do not matter for a given row.

The final two columns indicate whether providing or not providing the control action in the given context could cause a hazard. Note that the affirmative answers (indicating an unsafe control action) are just as important to document as the negative answers (indicating when control actions are believed or assumed to be safe). These assumptions are important because they influence the UCAs and requirements that are (or are not) defined. Although documenting assumptions is widely recognized as an important part of system engineering and many accidents have been linked to undocumented assumptions, very few techniques provide ways to systematically identify assumptions as they are made. Using a context table such as Table 6 can help make it clear what assumptions are being made and ensure that they are recorded so they can be reviewed or revisited if changes are later made to the system.

Comparing Table 6 to the initial set of unsafe control actions reveals a new unsafe control action. The third row describes a case where the range command does not match the driver selected range:

- **New UCA-8**: Shifter Control Module provides a range command that does not match the new range selection provided by the driver

  The context table can also help identify conflicts between the UCAs and identify multiple safety constraints that cannot all be satisfied. For example, rows 4 and 5 have conditions that overlap (i.e. the driver selected range matches SCM command and the selected range is not available) but they give conflicting answers: row 5 indicates that it is hazardous to provide the command while row 4 indicates it is hazardous not to provide the command. Table 7 shows the two UCAs and safety constraints side-by-side. This is an example of an unresolved conflict, and the Shifter Control Module as currently conceived may have no choice but to cause UCA-1 or UCA-4 through either action or inaction. In other words, in the current design it is not possible to satisfy both of the corresponding safety constraints if the driver selects an unavailable range.

  Once the conflicts and ambiguities are identified, there are often several potential ways to resolve them. In this case, one solution might be to design the Shift Control Module to take a special action when the driver selects an unavailable range (such as warning the driver or re-trying the range). The UCAs and safety constraints can then be updated accordingly. However, the most significant challenge is often not in ensuring that known problems are solved but in ensuring that unknown problems are identified early and will not go unnoticed until late in

TABLE 7 Unresolved conflict between two safety constraints

	<= Conflict =>	
UCA	UCA-1: Shifter Control Module does not provide range command when driver selects new range	UCA-4: Shift Control Module provides range command when that range is unavailable
Safety Constraint	SC-1: Shifter Control Module must provide range command when driver selects new range	SC-4: Shift Control Module must not provide range command when that range is unavailable

development or even during production. By formalizing the unsafe control actions early, the results can be used to drive the design and requirements in real time. In this case, SC-1 could be revised as follows:

- **Revised SC-1**: Shifter Control Module must provide range command when driver selects new range that is available

  Because the context table is based on a formal definition of an unsafe control action, conflicts may be automatically identified from the table using software tools. The completed context table also represents a model of the controller that issues the control actions and can be used to automatically generate executable safety requirements. For more information about automatic conflict detection and requirements generation, see [16]. For more information about tools being developed for this purpose, see [17].

3.7 Scenario Refinement and Adding Design Detail

At this point in the process, the causal accident scenarios that have been identified do not include much detail about the design. Some of the causal scenarios were already resolved without needing to dive into the design details. For example, S-2 in Table 5 was addressed by adding new feedback to the control structure. However, other scenarios may depend on more specific design details. For example, consider the following scenario:

- **S-3**: Shifter Control Module does not provide range command because it receives incorrect feedback that the range is already selected

  Without additional detail, it is not clear what could cause the range feedback to be incorrect or what controls are needed to prevent this scenario. To refine the scenario, it is necessary to "zoom in" on the Current Range feedback in the control structure and incorporate decisions such as:
- How are the range commands implemented?
- How is the current range sensed?
- What values can be reported to the Shift Control Module?

Notice that none of this information was needed until now, which is desirable because the answers may not have been available previously if development is occurring in parallel. However, once potential solutions are eventually proposed they can be immediately incorporated into the model by expanding the appropriate part of the control structure. Figure 7 shows a revised control structure with new detail shown in the dotted box. A new controller called the Range Motor Controller is added to monitor changes in the motor position, calculate the current range, and report the result to the Shift Control Module. In this example, it was also decided that the available range data could be obtained from an existing transmission controller, creating a new feedback loop. By refining the control structure and adding detail as it becomes available, the existing scenarios can be refined to identify more specific causes of unsafe control and determine whether the safety constraints are adequately enforced.

Given the additional detail in Figure 7, scenario S-3 can now be refined to include more detailed causes such as measurement errors or glitches that could result in a fixed offset when the relative movements are integrated. Notice that the current design does not yet have any way to detect this problem or to recover from it. As shown in Table 8,

FIGURE 7 Revised control structure with additional detail in the dotted box.

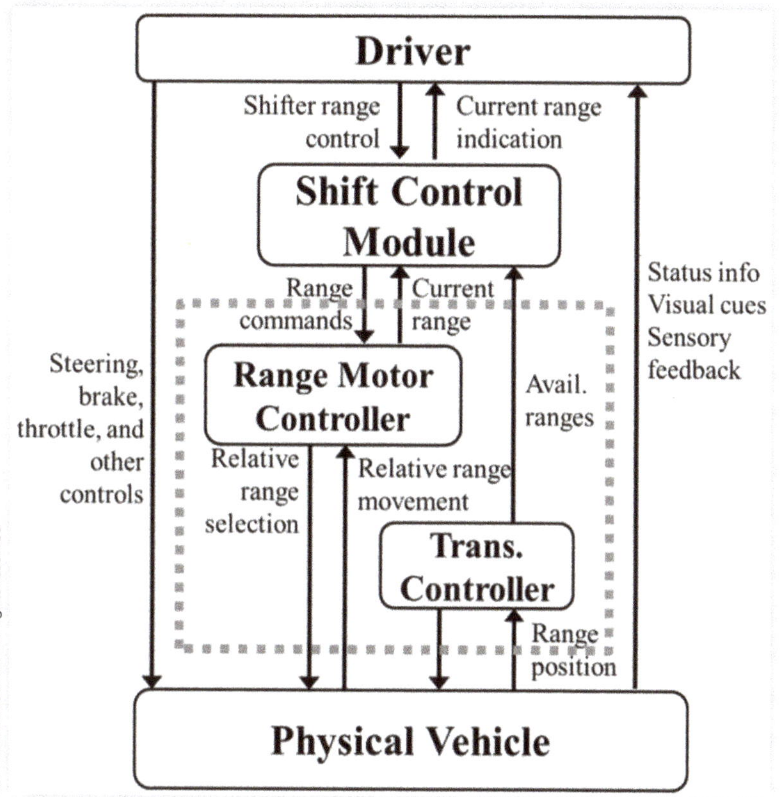

TABLE 8 Potential design solutions to prevent causal scenarios that violate safety constraints

Refined Safety Constraint	Detailed Causal Scenario that Violates the Constraint	Potential Controls or Design Decisions
SC-1: Shifter Control Module must provide range command when driver selects new range that is available	**S-3:** Shifter Control Module does not provide range command because it receives incorrect feedback that the range is already selected. The incorrect feedback could be caused by a fixed offset in the range motor controller due to a measurement error or a momentary glitch in the relative range movement signal.	- Provide a way to verify the current range (e.g. from transmission controller) - Provide a different type of feedback other than detecting relative movement - Provide a way to reset the transmission range (E.g. reset any time range is changed, provide automatic or manual reset capability, etc.)

multiple solutions are possible-use a different type of measurement, use the transmission controller to independently verify the current range, provide a way to manually reset the transmission range, etc.

This process can be repeated for any remaining scenarios that are not already controlled while scenarios and causes that are already prevented at a high level of abstraction do not need to be studied in more detail. In this way, the complexity of the analysis and the development process is managed while efficiency is improved by providing immediate feedback as the development progresses.

4. Discussion and Conclusion

The model-based safety-driven design process described in this paper integrates hazard analysis with the design process to build safety into a system as opposed to relying on after-the-fact or periodic analysis. The process was demonstrated on an initial shift-by-wire concept that replaces mechanical shift controls with electronic and computer controls. Intermediate results were used to define the necessary safety constraints/requirements, and corresponding design features were introduced as needed throughout the process. The process was also found to be more efficient because there was no need to wait for a completed analysis before making changes to the design, and the full analysis did not have to be repeated as design changes were made.

The first iteration can be performed very quickly and requires very little information to begin. The issues identified at this stage tend to be broad in nature, such as feedback or measurements that are altogether missing from the initial system. However, these issues can be easily addressed without requiring major rework when the process is started early. Most of the accident scenarios identified in the first iteration can be immediately addressed, but a few require more detailed study.

The second iteration is more rigorous and identifies deeper issues such as inconsistent safety constraints or feedback that exists but may not be trustworthy. A little more detail is required and the second iteration may take longer than the first, but the process is efficient and leverages all the results from the first iteration to avoid any wasted effort. The detailed scenarios that are identified at this stage do not require major design changes but instead tend to be governed by decisions naturally made during detailed design (such as the type of sensor used to detect the current range).

Some of the issues identified with this approach involve traditional component failures such as a failed sensor. However, many of the issues involved potential requirements and interaction problems that can be much easier to overlook. For example, Table 8 suggests that an additional control action and perhaps a new actuator should be added to the design concept. If the actuator had been included in the initial design then most traditional approaches could easily analyze the physical component, apply failure modes, etc. However few techniques can systematically identify components that are altogether *missing* from the design and had never even been conceived of. Similarly, the proposed approach has the ability to identify potentially hidden design assumptions unlike traditional analysis techniques.

Another important result of this approach is that behavioral software requirements were easily defined in Table 4 based on the unsafe control actions. Although it is fairly straightforward to derive hardware reliability requirements such as such as failure rates based on overall reliability objectives, other requirements can be much more challenging. In particular, there are very few methods that systematically produce the necessary software behavioral requirements such as "Software must provide X output whenever Y occurs". It is also very encouraging that missing requirements could be systematically identified using the proposed process (e.g. from UCA-8). Given the growing complexity of software requirements and the features they provide, this result could play an important role in reducing the number of problems discovered late and the amount of rework needed during later phases of development.

Although much of the focus has been on software, the process also considered important interactions between the vehicle and the human driver such as the driver selecting unavailable ranges. The driver was modeled just as easily as the software was-both are controllers that provide control actions based on beliefs about the system, and both use feedback to update an internal process model of the system. The causes of unsafe behavior, such as inadequate feedback and incorrect beliefs, are also very similar between the two and can be analyzed using the same process.

When the process is finished, executable software requirements can be generated and used to guide detailed software development, produce a controller model, perform model-based simulations, or intelligently generate test cases that are relevant for safety. Tools that can help produce executable requirements are being developed [17] and tools that can simulate the existing requirements are available [18, 19, 20].

References

1. Siemens, *Ford Motor Company Case Study* (Siemens PLM Software, 2014), Retrieved from http://www.plm.automation.siemens.com/pub/case-studies/14303?resourceId=14303.
2. McKendrick, J., *Cars Become 'datacenters on wheels', Carmakers become Software Companies* (ZDJNet, 2013).
3. NHTSA, "Office of Defect Investigation, Recalls [Data file]," 2014, Retrieved from http://www-odi.nhtsa.dot.gov/downloads/flatfiles.cfm/FLAT_RCL.zip.
4. Charette, R., "This Car Runs on Code," *IEEE Spectrum* (2009).
5. Leveson, N., *Safeware: System Safety and Computers* (Reading, MA: Addison-Wesley, 1995).
6. Lutz, R.R., "Analyzing Software Requirements Errors in Safety-Critical, Embedded Systems," *IEEE International Conference on Software Requirements*, (1992).
7. Leveson, N., "Role of Software in Spacecraft Accidents," *Journal of Spacecraft and Rockets* 41, no. 4 (2004): 564-575.
8. ISO 26262:2011, "Road Vehicles - Functional Safety," International Standardization Organization, November 2011.
9. International Council of Systems Engineering, "Handbook of Systems Engineering," V3.2.1, 2011.
10. IEC 61508, "Functional Safety of Electrical/Electronic/Programmable Electronic Safety-related Systems," International Electrotechnical Commission, Edition 2.0, 2010-04.
11. Van Eikema Hommes, Q., "Review and Assessment of the ISO 26262 Draft Road Vehicle - Functional Safety," SAE Technical Paper 2012-01-0025, 2012, doi:10.4271/2012-01-0025.
12. Leveson, N., *Engineering a Safer World* (Cambridge, MA: MIT Press, 2012).
13. Balgos, V.H., "A Systems Theoretic Application to Design for the Safety of Medical Diagnostic Devices," Master's thesis, MIT, 2012.
14. Torok, R. and Geddes, B., "Systems Theoretic Process Analysis (STPA) Applied to a Nuclear Power Plant Control System," *MITSTAMP Workshop*, March 2013.
15. Leveson, N., Wilkinson, C., Fleming, C., Thomas, J., and Tracy, I., "A Comparison of STPA and the ARP 4761 Safety Assessment Process," MIT PSAS Technical Report, 2014.
16. Thomas, J. "Extending and Automating a Systems-Theoretic Hazard Analysis for Requirements Generation and Analysis," Ph.D. dissertation, Engineering Systems Division, MIT, 2013.
17. Thomas, J. and Suo, D. "An STPA Tool," *3rd STAMP/STPA Conference*, Cambridge, MA, 2014.

18. Leveson, N., Heimdahl, M., and Reese, J. "Designing Specification Languages for Process Control Systems: Lessons Learned and Steps to the Future," *Proceedings of the 7th ACM SIGSOFT International Symposium on Foundations of Software Engineering*, Springer-Verlag, Toulouse, France, 1999, 127-145.
19. Leveson, N. "Completeness in Formal Specification Language Design for Process-Control Systems," *Proceedings of the Third Workshop on Formal Methods in Software Practice*, ACM, 2000, 75-87.
20. Bellagamba, L., *Systems Engineering and Architecting: Creating Formal Requirements* (CRC Press, 2012).

CHAPTER 8

Integration of Multiple Active Safety Systems Using STPA

Seth Placke, John Thomas, and Dajiang Suo
MIT

Automobiles are becoming ever more complex as advanced safety features are integrated into the vehicle platform. As the pace of integration and complexity of new features rises, it is becoming increasingly difficult for system engineers to assess the impact of new additions on vehicle safety and performance. In response to this challenge, a new approach for analyzing multiple control systems as an extension to the Systems Theoretic Process Analysis (STPA) framework has been developed. The new approach meets the growing need of system engineers to analyze integrated control systems, that may or may not have been developed in a coordinated manner, and assess them for safety and performance.

The new approach identifies unsafe combinations of control actions, from one or more control systems, that could lead to an accident. For example, independent controllers for Auto Hold, Engine Idle Stop, and Adaptive Cruise Control may interfere with each other in certain situations. This paper demonstrates a method to efficiently identify potential unsafe scenarios without requiring a complete enumeration or individual analysis of all possible scenarios. As a result, the approach is scalable to large systems with many controllers. In this paper, the method is demonstrated through a case study involving several driver assistance systems including advanced brake controls, advanced engine control, and advanced adaptive cruise control. Potential conflicts that would prohibit safe and successful operation are also efficiently identified, allowing engineers to develop suitable controls that prevent these conflicts.

CITATION: Placke, S., Thomas, J., and Suo, D., "Integration of Multiple Active Safety Systems Using STPA," SAE Technical Paper 2015-01-0277, 2015, doi:10.4271/2015-01-0277.

Introduction

Commercial vehicles have recently seen a rapid introduction of new software-controlled features, from parallel parking and lane keeping to Engine Idle Stop and advanced Adaptive Cruise Control systems. As new features are added and integrated with previous systems, the overall complexity increases substantially, especially in terms of new potential interactions that weren't possible previously. This, in turn, has made it much more difficult to analyze all potential interactions or to ensure the combined systems will not behave in potentially hazardous ways. Some have turned to Use Cases as a way to examine expected situations, define the appropriate system responses, and consider potential side-effects on other vehicle systems [1]. However, Use Cases are inadequate for developing complete requirements in complex systems. Use Cases are rarely complete in the sense of capturing every potential scenario, and they are also inefficient and require individual analysis of many different scenarios to ensure that certain behaviors will never happen. In fact, accidents usually occur in off-nominal cases, i.e., conditions where the assumptions made in the Use Case are incorrect.

Traditional analysis methods, such as Failure Modes and Effects Analysis (FMEA) and Fault Tree Analysis (FTA), focus on predicted component failures. However, the increased complexity of modern systems has changed the types of accidents we see today. More and more accidents are being caused not by individual components that have failed but by software that operated exactly as required and components that have not failed [2]. In fact, most software-related accidents are caused by flawed requirements rather than coding errors or component failures [3]. To address these problems, new analysis methods are needed to identify when the required behavior is flawed and to identify potentially unsafe interactions that may result when systems are integrated. Furthermore, if the solution is to be practical for modern complex systems, it must be efficient and cannot rely on manually examining every potential interaction individually.

One way to capture feature interactions is by defining a formal executable model and assessing preconditions and postconditions for various operations [4, 5, 6, 7]. These approaches can be automated and can be much more efficient than manually enumerating each individual interaction. However, they require that a formal model of the system exists and they assume that the necessary preconditions and postconditions are already known and correct. In practice this may not be the case, especially during early development phases when most safety decisions are made [8]. In addition, interactions with other system elements can be challenging because accurate formal models of non-software components, such as humans, may never exist.

Systems Theoretic Process Analysis (STPA)

Systems Theoretic Process Analysis (STPA) [9] is a hazard analysis technique that can identify a broad array of accident scenarios including those due to component failures, dysfunctional interactions, flawed requirements, and other causes. One of the strengths of STPA is the applicability to early development phases and ability to capture interactions between many different types of components including hardware, software, human operators, human managers, and other components. More detailed comparisons and evaluations of STPA can be found in [10, 11, 12, 13]. However, very little guidance has been developed to systematically identify undesirable interactions between multiple features [14, 15]. In this paper, a new process is demonstrated to systematically identify undesirable interactions between multiple features during an STPA analysis.

STPA begins by defining the system of interest and the system accidents, which are any unacceptable losses that the system must not experience. The system hazards that may lead to an accident are then identified and prioritized. A system hazard is a system state that will lead to an accident in a worst-case environment. The system being in a hazardous state does not guarantee that an accident will always occur, but the hazards should still be prevented and mitigated through the system design. The defined system accidents and system hazards provide a foundation for traceability throughout the analysis effort, so that findings and results from lower levels of analysis may be traced backed to their system-level impact.

The system is modeled in terms of a functional control structure comprised of hierarchical control loops which may have both social and technical components and can be represented at varying levels of abstraction. Figure 1 is an example of a simple, generic system control structure in which a process is controlled by automated and human controllers who act on the process through actuators and receive feedback information through sensors.

After the functional control structure has been modeled, the first step of STPA is to identify how the controllers may issue control actions in a manner that is potentially unsafe or otherwise inappropriate.

There are four ways in which a control action may be hazardous:

1. A control action required for safety is not issued.
2. A control action is issued when it is unsafe to do so.
3. A control action is issued too soon or too late.
4. A continuous control action is applied too long or stopped too soon.

Instances when these four types of Unsafe Control Action (UCA) may occur are often recorded in a table for easy reference. Once they have been identified, UCAs may be translated into safety constraints which are high level constraints on system behavior that must be enforced to ensure an accident does not occur. STPA then moves to a second step that identifies reasons that a UCA may be issued and ways in which a correct control action may be issued but not executed properly. For example, a controller may have an incorrect belief about the state of the system (i.e. an incorrect process model) that causes the controller to issue an unsafe control action. A control loop with guidewords to prompt the analysts, such as that shown in Figure 2, is used to ensure a thorough analysis [3].

Once identified, the causal factors leading to accidents may be used to write requirements for the system components. The results of the causal factor analysis may be used to eliminate hazardous scenarios or, if elimination is not possible, reduce their occurrence or their impact.

STPA is a top-down methodology capable of analyzing complex systems and capturing interactive effects between subsystems. Significant guidance is provided throughout the process to ensure the system in question is analyzed in a methodical manner. Because the method can be used at many levels of abstraction, STPA is useful beginning at the concept development stage and through the entire design process.

FIGURE 1 Generic control structure example.

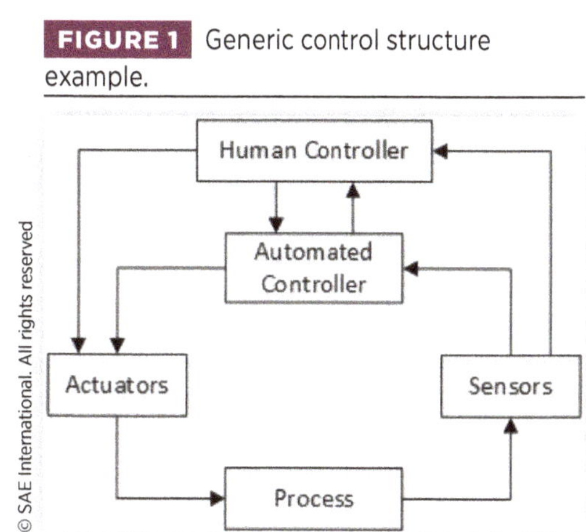

FIGURE 2 STPA Step 2 Guide Loop [3].

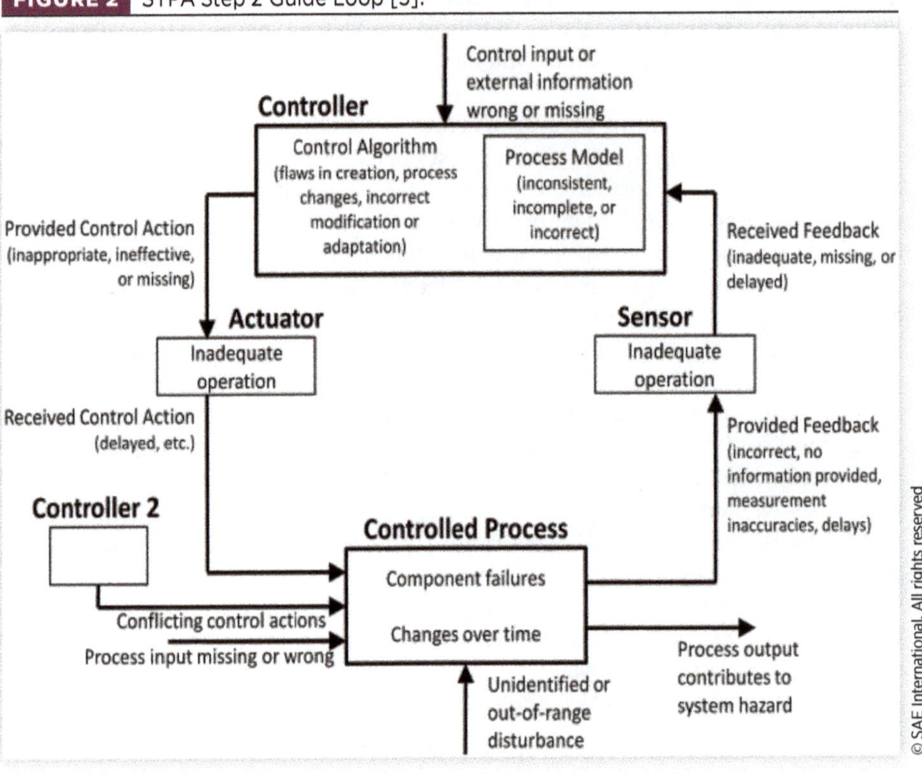

In this paper, additional guidance is developed to efficiently identify interactions between multiple controllers during STPA Step 1. Note that STPA Step 2 must still be performed to identify causes of unsafe control. The new process is demonstrated and evaluated using a case study that integrates three automotive features.

Case Study

The case study for this research involves three automotive features. Ideally, one would start with the highest level of system goals that, through decomposition, would eventually branch into features that can be developed by different engineering teams. However, it is often the case in industry that an engineer (or engineering team) is responsible for integrating legacy systems that were designed separately and it may not always be possible to start with a blank sheet of paper or develop the integrated system from scratch. The approach described in this paper can be applied in either case, however the case study below describes the more challenging case of three independently developed legacy systems that must be integrated safely.

Feature Descriptions

The three features in this case study are Auto-Hold, Engine Stop-Start, and Adaptive Cruise Control with Stop-Go. These features share a similar concept; an embedded controller is integrated into the vehicle to control existing hardware in the braking and propulsion systems in order to provide new and/or enhanced functionality. Due to proprietary considerations, the descriptions used in this paper are fictional and do not necessarily reflect the production intent of any manufacturers. However, these examples were developed in collaboration with professional automotive engineers to ensure that they are representative of what three independent features might look like before integration efforts and realistic in terms of the potential for undesirable interactions. Publicly available information from a number of manufacturers was also surveyed to ensure the examples are representative in terms of the types of features being introduced throughout the industry and their basic functionality.

AUTO-HOLD

Auto-Hold (AH) is an automatic braking feature that holds the vehicle and prevents rollback, allowing the driver to remove his foot from the brake pedal without vehicle movement whether it is on an incline or not. When the vehicle is brought to rest using the brakes, the AH feature will maintain the necessary brake pressure to keep the vehicle from moving by capturing the pressure already in the system. Once sufficient wheel torque (other than idle torque) is supplied to move the vehicle forward, the feature will disengage and release the braking system back to its normal state. The AH feature uses existing hardware components including Anti-Lock Braking System (ABS) and Electronic Parking Brake (EPB) components. Auto-Hold may issue the following four high-level commands:

- **HOLD:** When AH is ENABLED and the vehicle is brought to rest using the brake pedal, the HOLD command is issued to capture the existing brake pressure and place the feature in HOLD-MODE. AH identifies this situation by monitoring brake-pressure and wheel speed as provided by the braking system.
- **ADDITIONAL PRESSURE:** When the system is in HOLD-MODE and the wheels begin to rotate, the ADDITIONAL-PRESSURE command is issued to increase brake pressure (using the ABS pump) until the vehicle comes to rest.
- **RELEASE:** When the system is in HOLD-MODE and one of two conditions is met, RELEASE is issued to release the valve and return the brake system to normal operation. These two conditions are: 1) The propulsion torque is sufficient to move the vehicle. 2) Another system takes responsibility for holding the vehicle.
- **APPLY EPB:** When the AH system is in HOLD-MODE, it may engage the EPB. This control action may be issued if the hydraulic brakes are not effective.

ENGINE STOP-START

Engine Stop-Start (SS) is a feature designed to reduce fuel consumption and emissions by turning off an engine that would otherwise be idling while the vehicle is stopped. When the vehicle comes to a complete stop, the engine is automatically turned off and then restarted before motion resumes. The two high-level commands are:

- **STOP:** Once the vehicle is brought to rest using the brake pedal, *Auto-Stop* is issued which shuts down the engine.
- **RESTART:** When the system is in AUTO-STOPPED and power needs arise, RESTART is issued to restart the engine.

Adaptive Cruise Control with Stop-Go

Adaptive Cruise Control with StopGo (SG) is an enhanced version of the legacy cruise control feature. With traditional cruise control, a driver may pre-set a speed for the vehicle to maintain, allowing him to remove his foot from the accelerator pedal.

Traditional cruise-control systems allow the driver to set the speed, increment the set speed up or down, temporarily increase speed using the accelerator pedal (such as when passing), and disengage the feature using the driver controls or the brake pedal.

Adaptive Cruise Control (ACC) builds upon this traditional architecture in that it intelligently considers the distance to vehicles and other objects ahead of the primary vehicle in the same lane. Radar is mounted in the front of the vehicle that reports the range and range-rate to a controller that may utilize the braking and propulsion systems to maintain a safe trailing distance.* As with speed, the driver can change the desired trailing gap through the feature controls.

The Stop-Go capability further builds upon this architecture, enabling the ACC system to bring the vehicle to a full stop and resume motion when following a target vehicle. This is intended to be used in stop-and-go traffic when the vehicle may momentarily come to a stop before moving forward in concert with the surrounding vehicles.

- **ACCELERATE:** When the system is enabled, it may provide the accelerate command to accelerate the vehicle using the propulsion system.
- **DECELERATE:** When the system is enabled, the decelerate command will decelerate the vehicle using the braking system.

Potential for Unsafe Interaction

As mentioned previously, all three of these features have some control authority over the brake and/or propulsion subsystems and thus have the potential to interact through the vehicle dynamics. It is possible that the features may interact positively, for example if Auto-Hold maintains the vehicle's rest position thereby enabling the Engine Stop Start feature to halt fuel consumption while the vehicle is held stationary. However, it is also possible for the features to interact negatively, for example if the Auto-Hold feature prevents the ACC w/Stop-Go feature from taking off after a brief pause in traffic or if the Engine Stop-Start feature shuts off the engine while ACC w/Stop-Go is attempting to accelerate or resume.

Identifying potential instances of these types of negative interactions, both inconveniences and potential hazards, is the goal of this case study and the techniques demonstrated. We will show that STPA and the new technique can identify instances of potentially dysfunctional and/or hazardous interaction at the beginning of the engineering design process, before significant development work has occurred, and prevent significant rework during verification and validation testing.

Analysis

SYSTEM ACCIDENTS AND SYSTEM HAZARDS

Tables 1 and 2 present the sets of vehicle-level system accidents and system hazards that were defined at the beginning of the analysis. These system accidents represent the losses that should be avoided during operation and the system hazards represent the system states that might lead to such losses.

For example, A-1 could occur if a trailing vehicle rear-ends a leading vehicle in city traffic. A-2 could occur if a vehicle collides with a pedestrian or a guardrail. A-3 could occur if there is excessive deceleration resulting in whiplash.

* Trailing gap may be implemented as a time-gap to the leading vehicle so that it can vary dynamically with speed.

For example, H-1 could occur if a vehicle accelerates into another vehicle ahead. H-2 could occur if a vehicle experiences a near miss with a pedestrian or other object. H-3 could occur if a vehicle accelerates too fast for weather conditions. H-4 could occur if there is excessive temperature such as occupant heat exhaustion or burns from hot surfaces.

TABLE 1 System accidents

Number	System Accident Description
A-1	Two or more vehicles collide
A-2	Vehicle collides with other obstacle or terrain
A-3	Vehicle occupants injured without vehicle collision

TABLE 2 System hazards

Number	System Hazard Description	Accident
H-1	Vehicle does not maintain safe distance from nearby vehicles	A-1
H-2	Vehicle does not maintain safe distance from terrain and other obstacles	A-2
H-3	Vehicle enters uncontrollable/ unrecoverable state	A-1, A-2, A-3
H-4	Vehicle occupants exposed to harmful effects and/or health hazards	A-3

System Control Structure

Following the definition of system accidents and system hazards, the system control structure was developed as shown in Figure 3.

The system control structure is a functional representation of the system as hierarchical control loops. The control structure consists of functional blocks connected by arrows that represent control actions and feedback. When drawing the control structure, the analyst assigns responsibilities to each functional block and connects them

FIGURE 3 Combined system control structure.

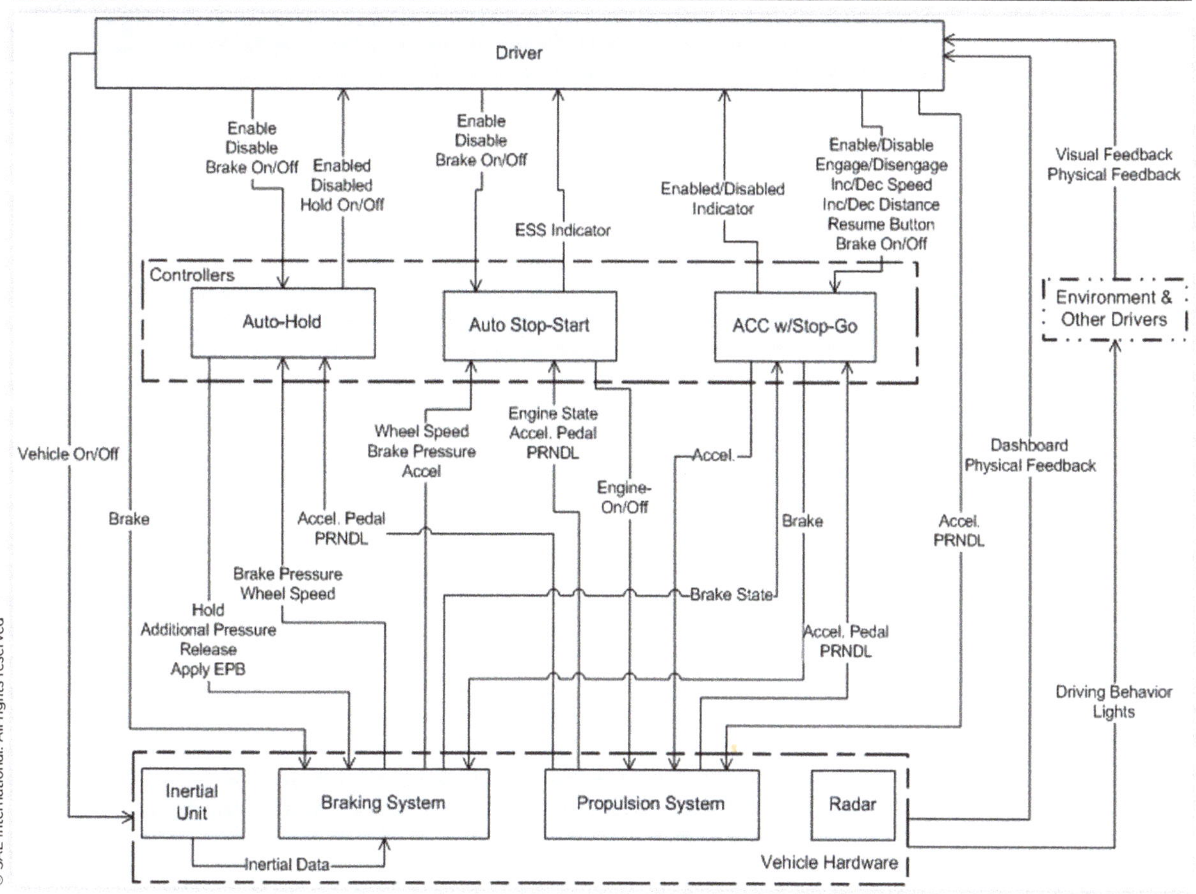

in a hierarchy. Defining a global control structure helps ensure that all system stakeholders have a common understanding of the system's design and operation.

The blocks in Figure 3 represent functional entities in the system. The convention used here is that entities higher on the structure have authority over entities lower on the structure and that control actions flow down while feedback flows up. For visual simplicity, only one arrow in a given direction is shown between two blocks-so an arrow represents a path rather than an individual command or piece of feedback. Multiple terms labeling an arrow represents the set of control actions or feedback variables (depending on direction) that may exist on the path indicated by the arrow.

The control structure shows that the three feature systems and the driver have overlapping control authority and receive feedback from common processes. This overlap and commonality creates the potential for interaction that may lead to unsafe system behavior.

Identifying Unsafe Control Actions

Figure 3 shows that each of the controllers in the system may issue commands, also referred to as control actions. As discussed previously, there are four ways in which the issuing of a control actions may be unsafe. These four manners of unsafe control must be considered for each control action issued in the system. In this system, the automated controllers issue a total of nine high-level control actions. For space considerations, analysis of only one of these control actions will be shown to demonstrate the process applied to the entire system.

When engaged, the Auto-Hold feature may issue the RELEASE command to return the braking system to normal operation. The circumstances in which it may be unsafe to issue the RELEASE command are listed in Table 3.

The UCA's listed in Table 3 may be further refined and formalized as shown in the Context Table presented as Table 4. This method of formalizing UCA's was developed and first presented by Thomas in [16]. The Thomas method rests on the premise that UCA can be formally defined by a set of Process Model Variables that define the state

TABLE 3 UCA Table for the Auto-Hold Command: RELEASE

Controller: Auto-Hold Module				
Control Action	**Not Provided**	**Provided**	**Too Soon/Late**	**Too Long/Short**
RELEASE	UCA-AH-11: Not providing RELEASE is hazardous if driver has commanded sufficient wheel torque via the accelerator pedal [H-1,2,3]	UCA-AH-12: Providing RELEASE is hazardous if AH is in HOLD-MODE and driver has not commanded sufficient wheel torque [H-1,3]	UCA-AH-13: Providing RELEASE before the there is sufficient wheel torque is hazardous [H-1,3]	
	UCA-AH-14: Not providing RELEASE is hazardous if the driver DISABLES AH [H-1,3]			

TABLE 4 Context Table for the Auto-Hold Command: RELEASE

Context ID	AH Enabled	Hold-Mode	Driver Present	PRNDL	Gas Pedal	Propulsion Torque Sufficient	Brake Pedal	EPB On	Providing Causes Hazard	Not Providing Causes Hazard	Providing Required for Function	Not Providing Required for Function
AH-R-1	Yes	Yes	Yes	P	*	*	*	No		x		
AH-R-2	Yes	Yes	Yes	N	*	*	Not Pressed	No	x			
AH-R-3	Yes	Yes	Yes	N	*	*	Pressed	No	x			
AH-R-4	Yes	Yes	Yes	R,D,L	Pressed	Yes	*	No		x		x
AH-R-5	Yes	Yes	Yes	R,D,L	Pressed	No	*	No	x			
AH-R-6	Yes	Yes	Yes	R,D,L	Not Pressed	*	*	No	x			
AH-R-7	Yes	Yes	No	*	*	*	*	No	x			
AH-R-8	No	Yes	*	*	*	*	*	No	x			
AH-R-9	*	No	*	*	*	*	*	No				
AH-R-10	*	*	*	*	*	*	*	Yes				

of the system as viewed by the controller. In other words, the Process Model Variables are used by controllers to determine the appropriate control actions. Each row in the table corresponds to a UCA and asterisks are used to denote which variables do not matter for a given row. In addition to providing increased formalism and clarity, the Thomas method enables automated identification of conflicts and the generation of executable requirements.

Identifying Conflicts between Controllers

One way that conflicts occur is when the effects of one command violate the assumptions and conditions of another. As a consequence, one or more controllers may be prevented from realizing their intended functions resulting in potential performance and safety issues.

Another way that conflicts occur is when the effects of one command satisfy the assumptions and conditions of another, triggering the second command to be issued at an unanticipated time or in an uncoordinated manner. This type of conflict is more subtle than the first that arises from commands effectively prohibiting each other. In this case, the requirements for an individual command to be safe and functional are still met; however, from a system perspective, the control amongst controllers is uncoordinated such that they may be competing against each other.

Creating a Conditions Table

To identify potentially unsafe interactions, a conditions table like the one in Table 5 is used. The conditions table is a table that keeps a record of the design assumptions, required conditions, and effect on the system associated with each command.

TABLE 5 Conditions table format

	Controller 1		Controller 2	Controller 3	
	Command A	Command B	Command C	Command D	Command E
Design Assumptions & Required Conditions					
Effect on the System					

The fields in the conditions table can be partially populated with the results of the UCA analysis presented earlier. The formal analysis of UCA's using Context Tables produces the "Required Conditions" input for the conditions table. The "Design Assumptions" may also arise during the UCA analysis or can be elicited from the engineering design team. The "Effect on the System" can be populated by considering the process model variables identified previously in STPA.

As an example, the combination of rows in Table 4 that are not marked as 'Providing Causes Hazard' but are marked as 'Not Providing Causes Hazard' or 'Not Providing Required for Function' can be used to help define the required conditions for the AH RELEASE control action. These are further refined by the design assumption that RELEASE will be issued when AH is holding the brakes (as defined in the context table) to produce the results in Table 6.

The Condition Tables for Auto-Hold, Engine Stop-Start, and Adaptive Cruise Control with Stop & Go are presented below. The generic table presented as Table 5 has been inverted and broken into three sections (one for each feature controller) for readability. Appending Tables 6, 7, 8 and inverting the axes will put the information in the same format as presented in Table 5.

TABLE 6 Auto-Hold portion of the Conditions Table

		Design Assumption & Conditions Required	Effect on the System
Auto-Hold	**HOLD (HOLD-MODE)**	**AH Enabled:** Yes **Wheels Rotating:** No **Brakes:** On **Driver Present:** Yes **Gas Pedal:** No **Range:** D | L	**Brakes:** Applied by AH
	RELEASE	**Brakes:** Applied by AH **Propulsion Torque:** Yes **AND Driver Present:** Yes **OR Vehicle Held:** Yes (i.e. **Range:** Park | **EPB:** Yes) **Range:** D | P	**Brakes:** Not Applied by AH
	ADDITIONAL-PRESSURE	**Brakes:** Applied by AH **Wheels Rotating:** Yes **Electrical Power:** Available (i.e. **Engine:** On | **Battery:** High) **Brake Pressure:** <Max	**Battery:** Reduce Charge **Brake Force:** Increased

TABLE 7 Engine Stop-Start portion of the Conditions Table

		Design Assumption & Conditions Required	Effect on the System
Engine Stop-Start	STOP (AUTO-STOPPED)	**ESS Enabled:** Yes **AUTO-STOPPED:** Yes **Vehicle Held:** Yes (i.e. **Brake:** On | **EPB:** Yes) **Restart Possible:** Yes (i.e. **Battery Charge:** High) **Driver Present:** Yes **Gas Pedal:** No **Auxiliary Power Needs:** Low **Range:** !=P,R,N	**Propulsion:** Off **Idle Torque:** No **Electrical Power:** Off **AUTO-STOPPED:** Yes
	RESTART	**Vehicle Held:** Yes (i.e. **Brakes:** On | **Range:** Park | **EPB:** Yes) **Wheels Rotating:** No **Restart Possible:** Yes (i.e. **Battery Charge:** High) **Driver Present:** Yes **Range:** !=P,R,N	**Propulsion:** On **Idle Torque:** Yes **Electrical Power:** On - power reduced ~2s **AUTO-STOPPED:** No

TABLE 8 Adaptive Cruise Control with Stop & Go portion of the Conditions Table

		Design Assumption & Conditions Required	Effect on the System
Adaptive Cruise Control with Stop-Go	Accelerate	**ACC Enabled & Engaged:** Yes **Vehicle Held:** No (i.e. **Brakes:** Not Applied & **EPB:** No) **Driver Present:** Yes **Driver Gas Pedal:** No **Range:** D *If* **Target Locked:** Yes *Then* **Distance >= Threshold:** Yes **Speed <= Threshold:** Yes **Propulsion Power:** Available (i.e. **Engine:** On)	**Gas:** Applied by ACC **Speed:** Increased **Wheels Rotating:** Yes
	Brake	**ACC Enabled & Engaged:** Yes **Driver Pedal Input:** No **Gas:** Off **Driver Present:** Yes **Range:** D **Hydraulic Power:** Available (i.e. **Engine:** On | **Battery:** High)	**Brakes:** Applied by ACC **Speed:** Decreased

Searching for Conflicts

The conditions table can be reviewed and searched for instances when the effects of one control action conflict with the assumptions and required conditions of another. The conditions table grows on the order $O(n)$ where n is the number of control actions that must be analyzed for interactions and conflicts. In other words the table grows linearly for each new control action that must be analyzed. A brute-force approach such as enumerating Use Cases can be thought of as analyzing and populating an $n \times n$ matrix (where n is the number of control actions), which exponentially grows on the order $O(n^2)$ by contrast and only captures pairwise interactions. In contrast, a conditions table like Table 5 is much more practical for large systems. Although the number of conflicts may inevitably grow as more control actions are included, a tool can also be used to automatically search a conditions table like Table 5 thereby ensuring that the manual effort by the user remains on the order $O(n)$. STPA Step 2 can then be performed on the resulting list of conflicting control actions to identify potential causes.

Results

By following this method, more than 60 conflicts were identified allowing potential controls to be developed. These conflicts are the result of design or requirements flaws, and many of them are not immediately obvious without a systematic technique to identify them. Many also involve assumptions that may have seemed reasonable for individual features but are not appropriate for an integrated system. Although only two scenarios are discussed here due to space limitations, a more complete demonstration is available in [13].

Conflict Scenario 1

The first example result is a conflict between Auto-Hold and Adaptive Cruise Control w/Stop-Go. When both features are installed on a vehicle, they can conflict due to operation of the EPB. The conflict was identified as:

Conflict 16: AH APPLY EPB prior to ACC w/SG ACCELERATE violates the constraint *Vehicle Held: No*

When both AH and SG are installed and enabled in a vehicle, the following scenario is possible that would lead to a dysfunctional and potentially hazardous situation:

- Driver engages ACC w/SG to maintain a selected speed and safe following distance.
- Traffic slows to a stop; SG slows the vehicle and holds it at rest.
- Once at a stop, AH engages (because the brakes were applied) and captures the existing pressure in the brake lines.
- Some stimulus (see below) triggers AH to engage the EPB.

Outcome: ACC w/SG cannot ACCELERATE because the vehicle is held by the EPB...

In this scenario, EPB is not released. Although AH can apply EPB, it does not have the ability to release EPB as a safety precaution (see Figure 3). AH issuing the APPLY EPB control action will change the 'Vehicle Held' state to 'Yes.' This violates the ACC w/SG requirement that 'Vehicle Held' be 'No.' If the vehicle is being held, the engine torque from the ACCELERATE command (e.g. when traffic clears) will oppose the force holding the vehicle which may prevent or delay forward motion or potentially damage systems like the parking brake. Recall that as a safety precaution the SG system does not have authority to disengage the EPB (see Figure 3).

There are several reasons that the AH feature may engage the EPB, one being the driver disabling the AH feature. Other potential reasons can be identified in STPA Step 2.

At first glance, this conflict appears to lie outside the safety domain; however, it may have safety implications when the driver needs to quickly move the vehicle (i.e. it is stopped in an intersection) but is unable to do so because he or she must first recognize the EPB is applied, and then disengage and take control from SG.

This conflict may be resolved; three potential strategies are described below:

1. Require that ACC w/SG monitor the state of the EPB and allow it to disengage when appropriate. Allowing ACC w/SG to disengage the EPB resolves the conflict, but adds complexity in that it must now be decided when it is appropriate for an automated system (ACC w/SG) to disengage the EPB.
2. Require that an issuing of the EPB turns other features 'off' and requires Driver intervention to disengage the EPB. This change is simpler than the first potential resolution and does not require giving automation the authority to disengage the EPB. However, not giving automated systems the ability to disengage the EPB means that each time it is issued, the Driver will be required to intervene. If the driver is required to intervene often, this strategy may reduce the transparency with which some features operate and thus reduce their value.
3. Require that AH does not engage when ACC w/SG is engaged because AH becomes redundant when ACC w/SG is holding the vehicle still. Note, this change does not prevent AH and ACC w/SG from using the same strategy and physical hardware to control the vehicle's brakes. A design solution is that ACC w/SG effectively 'implements' AH when it brings the vehicle to a stop; however, this should be done within the ACC w/SG logic and the standalone AH system should remain disengaged.

As this conflict exists at a fairly high-level of abstraction and concerns the overlap of authority between Auto-Hold and Adaptive Cruise Control w/Stop-Go, the third potential resolution is perhaps the best choice.

Conflict Scenario 2

The first conflict involved a pair of controllers. This second conflict scenario is an example of conflict between three controllers: Auto-Hold, Engine Stop-Start, and the Driver. The following conflicts were detected using the new approach:

Conflict 21: AH RELEASE while engine AUTO-STOPPED violates the AH required condition *Propulsion Torque: Yes*

Conflict 51: Driver SHIFT prior to ESS RESTART may violate the constraint that the Range is in a forward gear

Again consider a vehicle that has both AH and ESS enabled:

- Vehicle comes to a stop, both the AH and ESS features engage successfully.
  - Driver attempts to move the vehicle backward:
  - Driver shifts to Reverse
- Driver applies the Accelerator Pedal

Outcome: The vehicle is effectively stuck, because:

- ESS is prevented from restarting the engine by FMVSS 102 [17].
- AH cannot RELEASE because there is insufficient wheel torque.

The stopping and starting of a vehicle engine is partially regulated by Federal Motor Vehicle Safety Standard (FMVSS) 102. The older version of this standard prohibited the engine starter from operating while the transmission shift lever is in either the forward or reverse drive position. This proves to be a barrier for new technologies, including hybrid-electric vehicles and idle-stop systems. In response to developments in those technologies, an updated version of FMVSS 102 prohibits the engine from automatically stopping in reverse gear but allows it to restart if automatically stopped while in a forward gear (i.e. followed the driver shifting to reverse) [17]. The engine may not automatically restart in reverse when the service brake pedal is not applied and must restart automatically when it is applied. This means that in the scenario described above, the engine may not automatically restart until the driver applies the service brake: AH maintaining brake pressure is not sufficient.

With the engine off, the driver pressing the gas pedal does not produce any propulsion torque and AH may not issue the RELEASE command. Thus, the vehicle is effectively stuck until the driver moves his foot to the service brake and the engine restarts. Once this happens, restored engine power will produce engine torque and if the driver presses the gas pedal, AH will issue RELEASE. However, this sequence may take several seconds during which the vehicle is effectively stuck and the sequence of driver actions required for recovery may not be obvious.

Resolving this conflict is not as straightforward as the first example as it involves several layers of control logic and regulation. Design engineers must prioritize the hazards associated with various solutions and choose one that is acceptable to all stakeholders.

Summary/Conclusions

This paper introduced a new method to identify potentially hazardous interactions among software-intensive features during STPA Step 1. The method was demonstrated by applying it to a case study with three independently developed features that were to be integrated. A number of potentially hazardous interactions were systematically identified, including interactions caused by potentially flawed requirements. More important, although a large number of potential interactions were possible, the method was found to be scalable and did not require enumeration of all possible interactions. Instead, hazardous interactions and conflicts were efficiently identified using much smaller condition tables.

Although a number of formal methods utilize pre-and post-conditions to search for undesirable feature interactions, these methods require a formal model of the system and assume that the required conditions are already known or given. The proposed method was able to derive the necessary conditions systematically from the context tables of an STPA analysis. In addition, the method did not require a formal model of the system to begin the analysis. This is important because formal models typically do not exist during early system development and may never exist for non-software components such as humans. As a consequence, this approach was successfully applied beyond software components to identify dangerous human interactions with automated systems such as driver shift commands before an ESS restart.

Future work includes a case study of greater complexity with a set of real vehicle systems. An example with controllers at multiple levels in the vehicle control hierarchy should also be considered to accurately represent modern production vehicles.

The method in this paper has the potential to be partially automated and tools can be developed to search for these types of conflicts. An open-source software tool is currently being developed to partially automate the STPA process, especially the identification of unsafe control actions in STPA Step 1 [18]. Once potential unsafe actions are identified, safety requirements can be generated or existing requirements can be checked to verify that unsafe behaviors are prevented.

Definitions/Abbreviations

ACC - Adaptive Cruise Control
AH - Auto Hold
EPB - Electronic Parking Brake
ESS - Engine Stop Start
FMEA - Failure Modes and Effect Analysis
FTA - Fault Tree Analysis
SG - Stop-Go
STPA - System Theoretic Process Analysis

References

1. Alladi, V., Wei, J., and Ganesan, S., "Writing Better Real-Time System Requirements with Use Cases and Services," SAE Technical Paper 2005-01-1315, 2005, doi:10.4271/2005-01-1315.
2. Leveson, N., "Applying Systems Thinking to Analyze and Learn from Events," *Safety Science* 49, no. 1 (2010): 55-64.
3. Leveso, N., "A New Accident Model for Engineering Safer Systems," *Safety Science* 42, no. 4 (2004): 237-270.
4. Tsui, F., Karam, O., and Bernal, B., *Essentials of Software Engineering* (Jones & Bartlett Learning LLC, 2014).
5. Hayes, I.J., "VDM and Z: A Comparative Case Study," *Formal Aspects of Computing* 4, no. 1 (1992): 76-99.
6. Zave P., "A Practical Comparison of Alloy and Spin," Formal Aspects of Computing, 2014, Available at Springer via http://dx.doi.org/10.1007/s00165-014-0302-2.
7. Leavens Gary, T. and Baker Albert, L., *Enhancing the Pre-and Postcondition Technique for More Expressive Specifications: FM'99 - Formal Methods* (Berlin, Heidelberg: Springer, 1999), 1087-1106.
8. Frola, F.R. and Miller, C.O., *System Sfety in Aircraft Acquisition* (Bethesda, MD: Logistics Management Institute, 1984).
9. Leveson, N., *Engineering a Safer World* (MIT Press, 2012).
10. Balgos, V.H., "A Systems Theoretic Application to Design for the Safety of Medical Diagnostic Devices," Master's thesis, MIT, Cambridge, 2012.
11. Torok, R. and Geddes, B., "Systems Theoretic Process Analysis(STPA) Applied to a Nuclear Power Plant Control System," Presentation at *MIT STAMP Workshop*, March 2013.

12. Leveson, N., Wilkinson, C., Fleming, C., Thomas, J., and Tracy, I., "A Comparison of STPA and the ARP 4761 Safety Assessment Process," MIT PSAS Technical Report, 2014.
13. Placke, S., "Application of STPA to the Integration of Multiple Control Systems: A Case Study and New Approach," Master's thesis, Engineering Systems Division, Massachusetts Institute of Technology, 2014.
14. Ishimatsu, T., Leveson, N., Fleming, C., Katahira, M., Miyamoto, Y., and Nakao, H., "Multiple Controller Contributions to Hazards," Presented at the *5th IAASS Conference*, Versailles, France, October 2011.
15. Ishimatsu T., Leveson N., Thomas J., Fleming C., Katahira M., Miyamoto Y., Ujiie R., Nakao H., and Hoshino N., "Hazard Analysis of Complex Spacecraft Using Systems-Theoretic Process Analysis," *Journal of Spacecraft and Rockets* 51, no. 2 (2014): 509-522.
16. Thomas J., "Extending and Automating a Systems- Theoretic Hazard Analysis for Requirements Generation and Analysis," Ph.D. dissertation, Engineering Systems Division, Massachusetts Institute of Technology, 2013.
17. "Federal Motor Vehicle Safety Standards; Transmission Shift Position Sequence, Starter Interlock, and Transmission Braking Effect," 49 CFR Part 571, 2005.
18. Thomas, J. and Suo, D., "An STPA Tool," Presented at *3rd STAMP/STPA Conference*, MIT, Cambridge, MA, 2014.

Integrating STPA into ISO 26262 Process for Requirement Development

Dajiang Suo, Sarra Yako, Mathew Boesch, and Kyle Post
Ford Motor Company

Developing requirements for automotive electric/electronic systems is challenging, as those systems become increasingly software-intensive. Designs must account for unintended interactions among software features, combined with unforeseen environmental factors. In addition, engineers have to iteratively make architectural tradeoffs and assign responsibilities to each component in the system to accommodate new safety requirements as they are revealed. ISO 26262 is an industry standard for the functional safety of automotive electric/electronic systems. It specifies various processes and procedures for ensuring functional safety, but does not limit the methods that can be used for hazard and safety analysis. System Theoretic Process Analysis (STPA) is a new technique for hazard analysis, in the sense that hazards are caused by unsafe interactions between components (including humans) as well as component failures and faults. Otherwise stated, STPA covers the safety analysis of system malfunctions as well as the safety of the intended function (SOTIF), in addition to Functional Safety.

This paper introduces a process map with a complete meta-model based on Systems Model Language (SysML) to support the integration of STPA into the functional safety process based on ISO 26262. In particular, the

CITATION: Suo, D., Yako, S., Boesch, M., and Post, K., "Integrating STPA into ISO 26262 Process for Requirement Development," SAE Technical Paper 2017-01-0058, 2017, doi:10.4271/2017-01-0058.

paper illustrates how STPA can help evaluate safety and other system-level goals with ASIL classifications from ISO26262's recommended Hazard Analysis and Risk Assessment (HARA). The meta-model can be also used to provide guidance on making architectural decisions in order to create functional safety requirements. To make the process map applicable to different functional safety processes adopted by OEMs, tool support is required. Guidelines on how to develop visualization tools based on the meta-model are given.

Introduction

Developing requirements for automotive electric/electronic systems is challenging for several reasons. First, engineers have to deal with not only safety-related goals early in the concept phase, but also other system-level goals such as performance and security that decide stakeholders' satisfaction with new products. Often, architectural tradeoffs have to be made before creating detailed requirements [1]. Second, traditional hazard analysis techniques that deal with hardware failures are hard to use in complex software-intensive systems [3, 15]. Unintended interactions among software features contribute to off-nominal scenarios even if the system operates as designed. Third, the modeling and tool support for system engineering activities often vary from department to department, making system integration difficult in the sense that models and requirement artifacts differ among various processes.

To deal with these challenges, theInternational Organization for Standardization (ISO) extended the general functional safety standard IEC 61508 to a domain-specific standard of functional safety for the automotive industry-ISO 26262 [2]. It specifies various processes and procedures for ensuring functional safety, but does not limit the methods that can be used for hazard and safety analysis. In particular, its concept phase prescribes the process for "identifying a comprehensive list of hazards and causal factors in order to support the development of safety requirements" [4]. STPA is a new technique for hazard analysis in the sense that hazards are caused by unsafe interactions between components (including humans) as well as component failures and faults. Previous work on the use of STPA in the automotive domain includes the application of STPA for identifying hazards in an ISO 26262 compliant framework and the corresponding modeling and tool support.

Hommes [4] suggests that STPA be used in the concept phase in ISO 26262 with other hazard analysis techniques for identifying a comprehensive list of hazards and developing functional safety requirements. Mallya et al. [5] shows how "STPA can be used in an ISO 26262 compliant process" in which hazard analyses based on STPA are extended to include risk assessment. Thomas et al. [6, 7] propose an integrated approach in which hazard analysis and requirement generation can be conducted in parallel and iterative process. Although not specific to ISO 26262, this approach provides guidance and feedback for engineers to make critical design decisions during early concept development.

For the modeling and tool support for hazard analysis and requirement generation based on STPA, software prototypes and tools have been developed. Suo [13] illustrates a proof of concept tool for supporting formalized STPA Step-1 and automatic requirement generation. Abdulkhaleq develops an open-source platform [8] that allows engineers to perform STPA for requirement development and test case generation. The Safety Hazard Analysis Tool (SafetyHAT) [9] is developed by Volpe, the National Transportation Systems Center, to support STPA and customization for applications in different transportation systems. Although these efforts facilitate the use of STPA in the automotive industry, to the best of the authors' knowledge, modeling and tool support are still lacking in the following three aspects:

- Augmenting STPA to consider safety as well as other system-level properties such as customers' experience or cyber-security threats early in the concept development.
- Verifying functional safety requirements by detecting conflicts between safety constraints automatically.
- Providing guidance for assigning responsibilities to system components for defining preliminary architecture.

FIGURE 1 An overview of the integration work.

Process Standard	ISO 26262
Analysis & Assessment method	HARA STPA
Modeling Language	System Modeling Language
Tool interoperation	XMI (other interchange format)

This paper introduces a process map for integrating STPA into the functional safety process defined by ISO 26262. In particular, it illustrates how to use STPA as the Hazard Analysis (HA) activity in an ISO 26262 compliant process for establishing safety and other system-level goals and to couple these to ASIL classifications that are based on current Risk Assessment (RA) methods to complete the HARA objectives of ISO 26262 Part 3. The process map described herein also provides guidance on making architectural decisions in order to accommodate functional safety requirements. To make the process map applicable to different functional safety processes adopted by OEMs, STPA extensions to the Systems Model Language (SysML) have been defined to facilitate using STPA with modeling and support tools. Guidelines on how to use the proposed process map are illustrated through a case study on an automated driving system. Figure 1 gives an overview of the integration work related to hazard analysis techniques in ISO 26262. All the work is based on ISO 26262-"Functional Safety Standard for Automotive Electric/Electronic systems" [2].

Process Map for SYPA Integration

ISO 26262

ISO 26262 is the standard for functional safety of electrical and electronic systems in vehicles and an adaptation of functional safety standard IEC 61508 [20] specific to the automotive industry [2]. It "prescribes a system engineering process for safety engineering" [18].

Three clauses in the concept phase of ISO 26262 that are related to the development of functional safety requirements are considered for integrating STPA [16]. Although revisions have been made since the publication of the preliminary standard in 2011, reviews and assessments of ISO 26262 in [16] are still valid. Processes for developing technical safety requirements are omitted as they involve design details of components.

- Item definition: A system or subsystem that achieves a function at the vehicle-level.
- Hazard Analysis and Risk Assessment (HARA): HARA is used to identify hazards and appropriate reduction of risk captured by safety goals. Engineers first

brainstorm possible hazardous events by deciding whether malfunctioning behaviors of a given item can cause hazards under various environmental conditions. The severity (S), exposure (E) and controllability (C) of the related hazardous event are assessed. Automotive Safety Integrity Level (ASIL) classification (A-D) is then assigned to each safety goal based on a fixed table mapping of {S, E, and C} to ASIL. For each hazardous event rated A-D, a safety goal (SG) is developed.

- Functional Safety Concept: The purpose of this step is to create functional safety requirements, at the concept level, to achieve safety goals developed in the HARA.

Although it gives a structuralized procedure for identifying hazardous events, the ISO 26262 standard provides limited guidance on specific hazard classification and mitigation.

STPA

STPA [3] is a new hazard analysis technique for complex systems in the sense that hazards are caused by unsafe interactions between components (including humans) as well as component failures and faults. STPA is based on STAMP-an accident causality models based on system and control theory [10]. Safety constraints, hierarchical control structure and process models are three basic concepts in STAMP. Accidents occur because safety constraints are not enforced by the control structure. Each "controller," whether human or computer, has a process or mental model for deciding whether or not to issue a control command to the process being controlled.

There are two steps in STPA, as shown in Figure 2. As an extension to Step 1, undesired control actions (UCA)* are identified and classified into four types:

- A control action required is not provided or not followed.
- An undesired control action is provided.
- A control action is provided too early or too late.
- A control action is stopped too soon or applied too late.

FIGURE 2 Procedures for system theoretic process analysis.

System Engineering Foundations:
- Hazards and losses
- Control structure
- System-level assumptions

STPA Step-1:
- Undesired control actions
- Safety Constraints

STPA Step-2:
- Scenarios
- Causal factors for UCAs

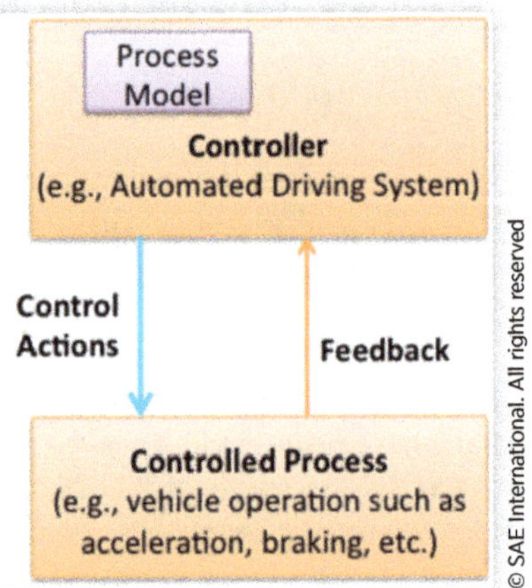

* UCA described in this paper is different from the standard definition [3] as other high-level goals of the system are considered.

After UCAs are identified, Step 2 identifies causal factors and scenarios that potentially lead to UCAs. These are based upon the accident causality model described in STAMP.

It is worth mentioning that STPA can be extended beyond safety to other high-level system goals such as security [17] or performance.

Process Map for Creating Functional Safety Requirement

This paper proposes a process map that couples STPA with the functional safety process in ISO 26262, as shown in Figure 3. The map shows how meta-models (center) described by System Modeling Language (SysML) can facilitate the use of STPA (right) in the concept phase (left) in ISO 26262 for developing functional safety requirements and making architectural decisions.

The meta-model (top center) first takes as inputs hazard information as well as other system-level properties such as customer experience that are used to identify undesired control actions in STPA Step-1, as indicated by the red arrow (downward) in Figure 3. Also, new security threats can be derived as system-level properties if there exists cyber vulnerabilities in causal factors found in STPA Step-2, as indicated by the upward (red) arrow in Figure 3. The meta-model can also provide assistance for deriving the control

FIGURE 3 Process map for integrating STPA with ISO 26262 process.

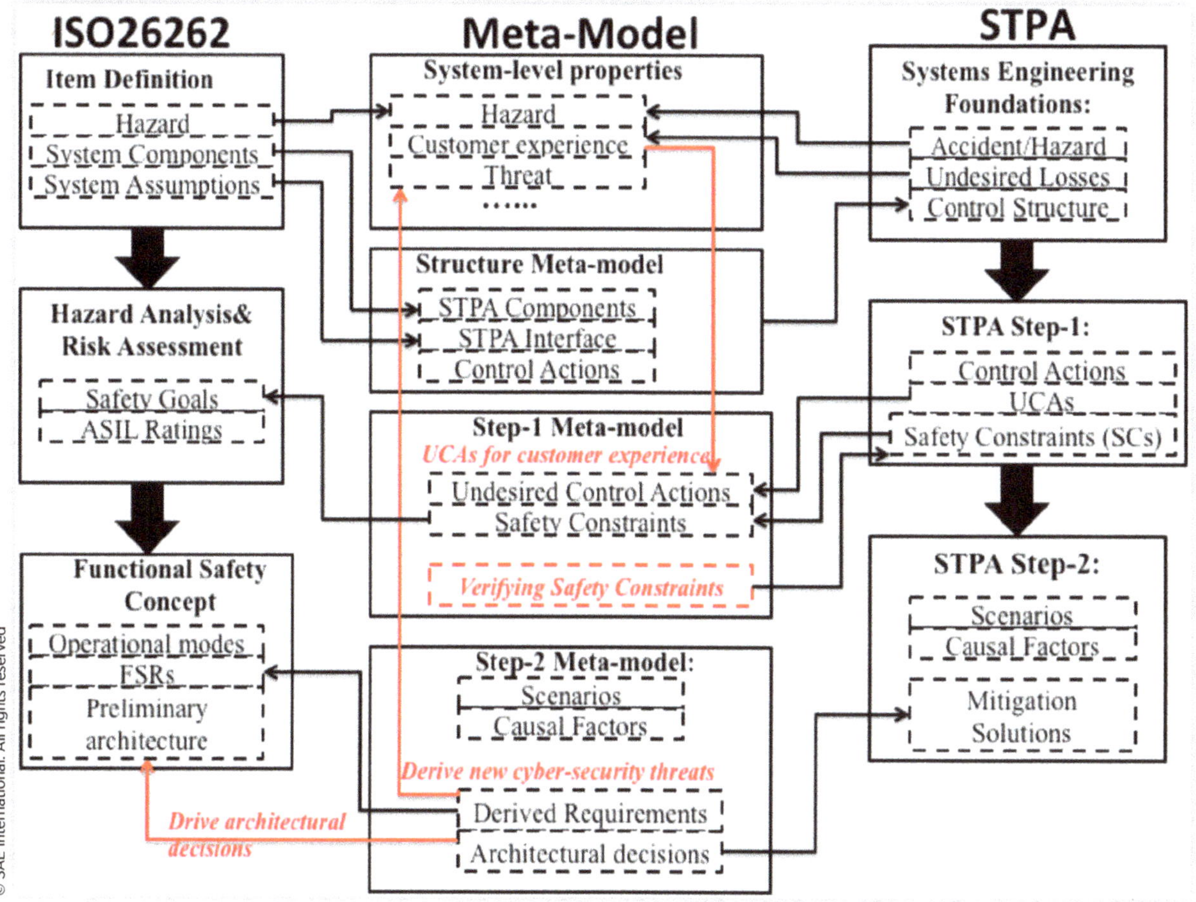

structure as systems engineering foundations based on system components and assumptions from the item definition in ISO 26262. The second role of the meta-model is to STPA to complement HARA. With tool support for STPA Step-1 based on the meta-model, engineers can create and verify safety constraints to be associated with ASIL rated safety goals from risk assessments. The third use of the meta-model is to provide guidance for creating functional safety requirements that drive architectural decisions, as shown at the bottom center in Figure 3.

Modeling and Tool Support

Intro to SysML

The OMG System Modeling Language (SysML) is a graphical modeling language that provides support for the development of complex systems including specification, analysis, design and verification [11]. It includes three types of diagrams, each of which represents a specific aspect of the system, including behavior diagram, requirement diagram and structure diagram. A block is the basic unit that structurally describes hardware, software, facilities, personnel and external entities in the environment in the structure diagram. For this project, SysML is used as the base language that was extended to capture the control structure for STPA. The SysML requirement diagram at the highest level relates Safety Goals to safety requirements and other non-safety requirements derived from analysis based on STPA.

Meta-Model for Hazard Analysis & Requirement Generation

As shown in Figure 1 and Figure 3, modeling support (from either open-source or commercial tools) is necessary to make the process map scalable to complex automotive systems and usable by different engineering teams. This paper describes a meta-model based on a modeling environment that supports Systems Modeling Language so that safety constraints and requirements derived in STPA can be mapped into the elements that support functional safety processes in ISO 26262, as shown in Figure 4, Figure 5 and Figure 6.

Figure 4 illustrates how a control structure (right) can be built by using the meta-model (left) that defines system components and links. In addition to components in STPA, new elements are added into the model-model for visualization. For example, the "Control Action" block travels on the "command" link, and thus an association link connects them.

Figure 5 (left) gives the meta-model that can couple hazard analyses based on STPA with the process in ISO 26262. The upper part shows how visualization tools (in table forms) can be built in order to associate UCAs and safety constraints from STPA Step-1 results with safety goals and ASIL ratings developed in HARA. Engineers can then leverage the traceability between UCAs and scenarios and causal factors to create corresponding functional safety constraints, as shown on the bottom.

In addition to modeling support for STPA integration with ISO 26262, this paper also proposes a meta-model for verifying safety requirements and constraints created

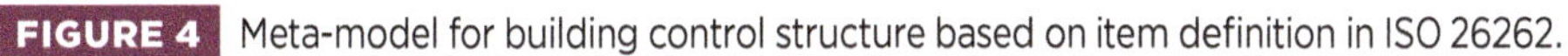

FIGURE 4 Meta-model for building control structure based on item definition in ISO 26262.

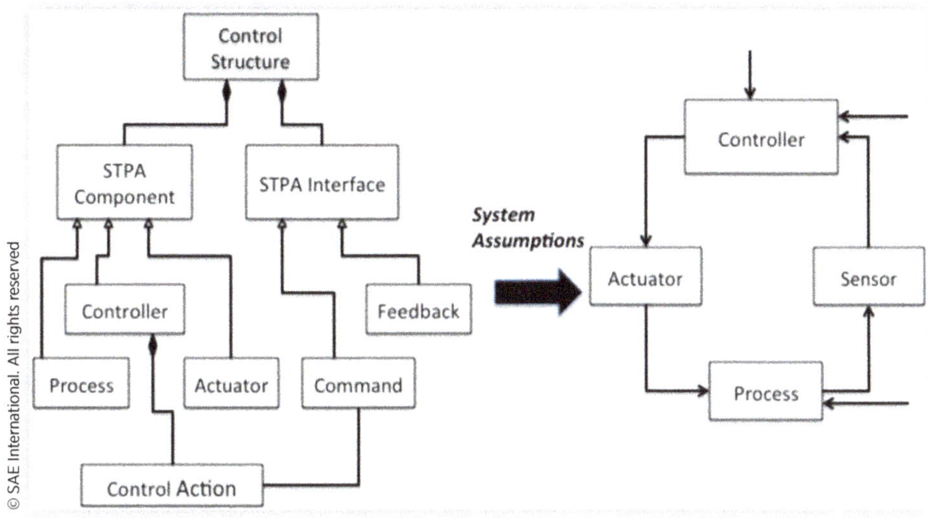

FIGURE 5 Meta-Model for supporting STPA Step-1 and Step-2.

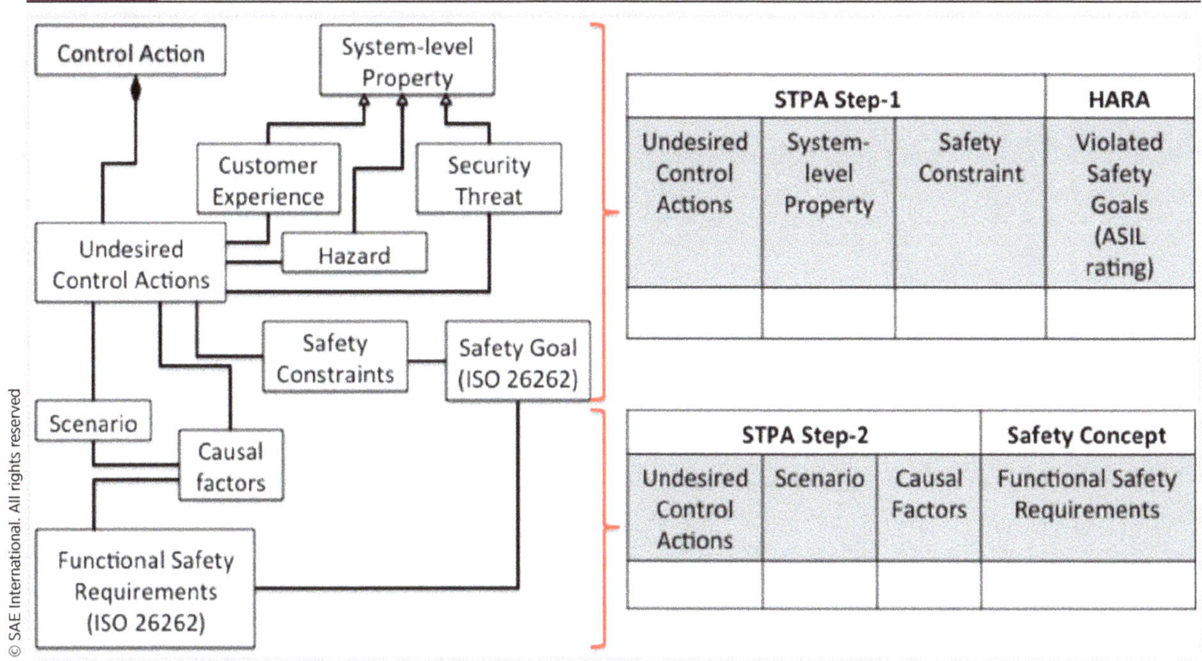

in the functional safety process, as shown in Figure 6. This meta-model builds on two theoretical foundations:

- Process model. A control structure in STPA can have multiple controllers (e.g., automated or human controller) and each controller has a model of the process being controlled and the external environment [10], as shown in Figure 2. A process model may contain different variables that represent the system states or context. For example, the automated driving system can be in active mode, within the operational boundary, and on the correct route. Whether or not it is desired to provide the motion planning command is dependent on this context.

FIGURE 6 Meta-Model for verifying safety constraints.

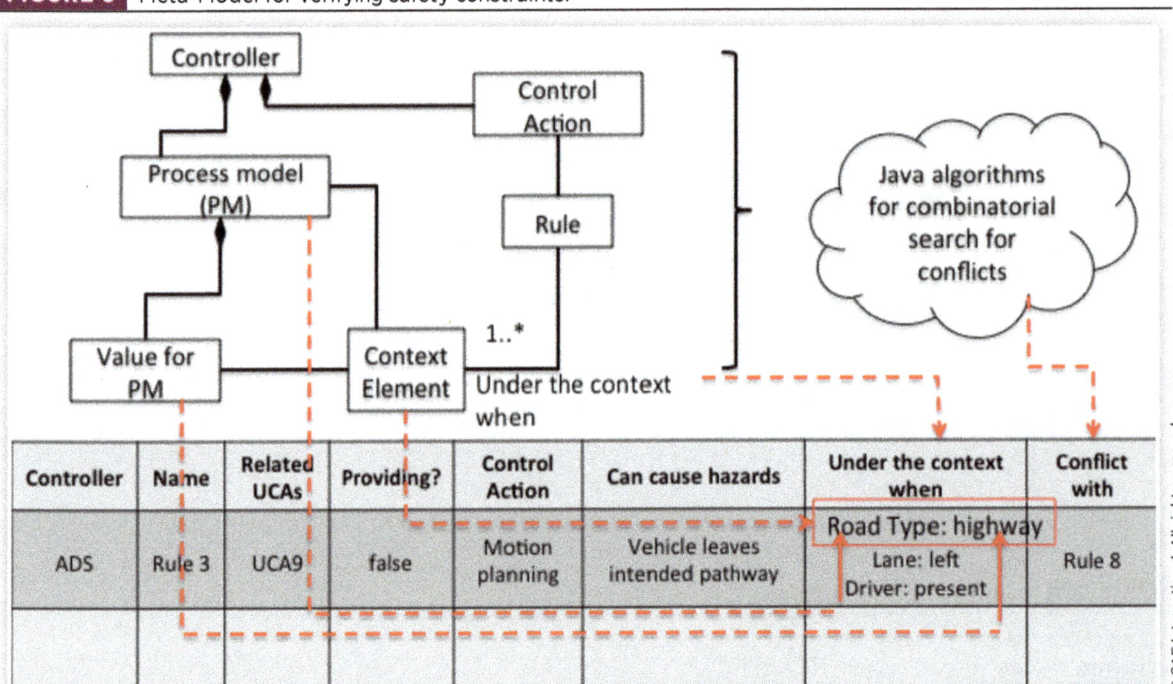

- A formal specification of hazardous control actions. Thomas [12] proposed a formal method to partially automate the process of identifying UCAs, identifying conflicts, and generating requirements. Specifically, a hazardous control action consists of four elements, including source controller, type (i.e., provided or not provided), control action and a context. With these four elements, engineers can define a "rule" which is the formalized UCA that contains a specific context. The rule is in a form that is close to natural language but machine readable for conducting combinatorial searches for determination of conflicting constraints.

Using the user interface based on the proposed meta-model in Figure 6, engineers can specify a "Rule" [12, 13] for UCAs. A "Rule" is a formalized way to specify contexts of a UCA [13]. The built-in Java algorithms in the plug-in developed by the author [13] can take these "Rules" as inputs and detect conflicts automatically. As shown in the last column of the table in Figure 6, "Rule 3" conflicts with "Rule 8", indicates a conflict between UCAs and corresponding safety constraints. A concrete example will be provided later in this paper.

System Engineering Foundations Based on Item Definition

To build the foundations for the system engineering process in ISO 26262, engineers take the inputs from item definition to construct a safety control structure with the modeling and tool support described in this paper. Those inputs include system-level hazards, system components and system assumptions.

As an example, three high-level hazards are chosen for this case study. G-1, a system-level goal related to customers' satisfaction or efficiency of the transportation system, is

FIGURE 7 Safety control structure of the automated vehicle derived from item definition.

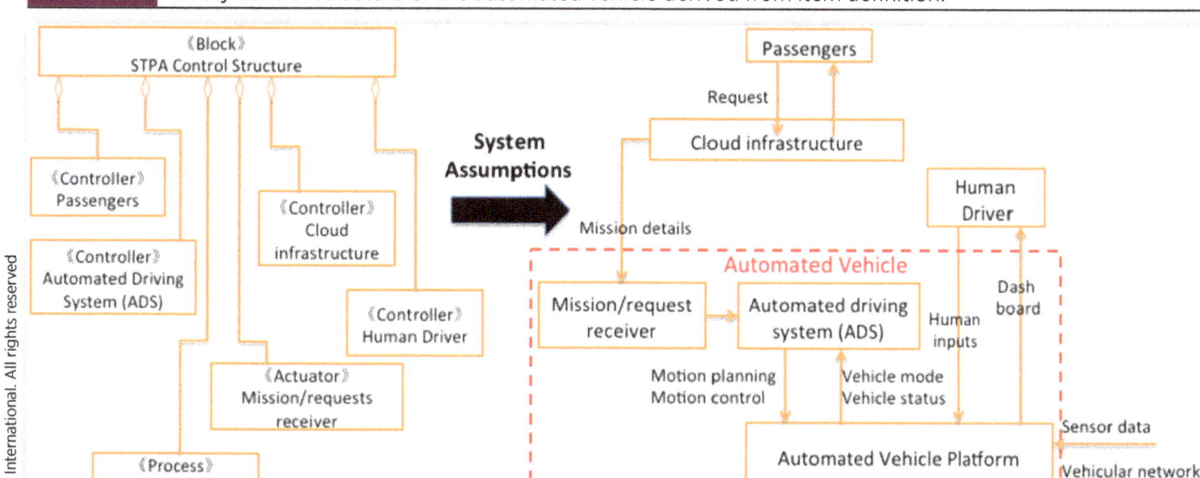

also included for illustrating the modeling support for dealing with different system-level properties.

- H-1: Getting too close to objects/terrain
- H-2: Vehicle leaves intended pathway
- H-3: Ingress/egress issue for rider
- G-1: Passengers' experience and traffic flow should not be disrupted by the operation of the automated driving system.

In addition, five assumptions are made on the automated vehicle:

- Passengers can request a ride service through the cloud infrastructure.
- The automated driving system and the road infrastructure support V2V or V2I communication through vehicular networks (e.g., Dedicated Short Range Communication - DSRC).
- The automated driving system can only operate within a prescribed operational boundary (e.g., time of the day, geographical boundary, weather conditions, etc.).
- After being activated, the automated driving system can handle all driving tasks appropriately.
- The human driver or passenger is able to interrupt the operation of the automated driving system.

The Modeling support for building system engineering foundations is shown in Figure 7. The SysML-based block diagram that represents system components (left) can be used to derive the control structure of the automated vehicle (right). This forms the basis of the example analyzed in the next section.

Integration of STPA Step 1 for Evaluating Existing Safety Goals

In STPA Step1, engineers identify undesired control actions to create initial (safety) constraints for the system. These constraints are then associated with existing Safety

Goals from the HARA. If any Safety Constraint is not associated with an existing Safety Goal, then the HARA can be revised resulting in a new Safety Goal.

Several abstract safety-goals developed from HARA are as follows.

- SG-1: Prevent vehicle from leaving intended pathway.
- SG-2: Prevent vehicle from violating fixed separation distance from other vehicles.
- SG-3: Prevent human ingress/egress issues.

Two control actions-acceleration and motion planning-are chosen for the case study. Consider UCA 9 below. If not providing "motion planning" can cause any of the hazards defined above, a safety constraint should be created.

- UCA 9: Not providing motion planning is undesired if the automated driving system is in active mode, the vehicle is in route and within operational boundary [H-1, H-2].
- Safety Constraints 9: The automated driving system should provide motion planning if is in active mode, the vehicle is in route and within operational boundary.

Table 1 shows how safety goals with ASIL ratings from HARA can be assigned to each UCA and its corresponding safety constraint. The first column includes a set of UCAs for acceleration and motion planning commands, while the third column shows safety constraints derived for each UCA. For a given row, if any Safety Constraint is not associated with an existing Safety Goal, then the HARA can be revised, resulting in a new Safety Goal.

As an example of how STPA can be extended to deal with non-safety goals, consider a situation where the automated driving system does not provide the acceleration command (the 6th row in Table 1) when the automated vehicle is merging into a fast (left) lane on the highway, and its current speed is too low to smoothly merge into the traffic flow.

TABLE 1 Assigning ASIL rated safety goals to safety constraints

STPA Step-1			HARA	
Undesired Control Actions	System-level Property	(Safety) Constraints	Violated Safety Goals (derived from HARA)	ASIL (derived from HARA)
UCA1: ADS Providing Acceleration command is undesired if the distance to the lead vehicle is less than the predefined threshold	H-1	SC1: ADS must not provide Acceleration command when the distance to the lead vehicle is less than the predefined threshold.	Prevent vehicle from violating fixed separation distance from vehicles	B
......				
UCA9: ADS Not providing motion planning command is undesired if the ADS is in active mode, the vehicle is en route and within operational boundary	H-1, H-2	SC9: ADS must provide motion planning command when in active mode, the vehicle is en route and within operational boundary.	Prevent vehicle from leaving intended pathway Prevent vehicle from violating fixed separation distance from vehicles	B B
Does Providing or Not Providing Acceleration or motion planning command violates G-1?	G-1			
New UCA derived from G-1				
UCAX: ADS Not providing Acceleration command when vehicle is merging into the fast (left) lane on the highway	G-1	Non-safety related	Non-safety related	N/A

TABLE 2 Detecting requirement conflicts automatically for verifying safety constraints

#	Name	Related UCAs	Providing?	Control action	Can cause hazards	Under the context when	Conflict with
1	Rule 3	UCA9	false	Motion planning	Vehicle leaves intended pathway	Road Type: highway Lane: left Human Driver: present	Rule 8
2	Rule 1	UCA10	true	Motion planning	Ingress/egress issue for rider	Road Type: City street Lane: left Human Driver: present Weather: snow	
3	Rule 2	UCA10	true	Motion planning	Vehicle is operating without human driver	Time: midnight Human Driver: not present	
4	......	......		Motion planning			
5	Rule 8	UCA10	true	Motion planning	Getting too close to objects	Road Type: highway Weather: storm	Rule 3
	......	......		......			

This is undesired (although not necessarily unsafe) because vehicles may need to brake and cannot maintain their desired speed that their passengers expect, thus violating G-1. Table 1 (last row) gives the newly derived UCA-ADS not providing Acceleration command is undesired when the vehicle is merging into the fast lane on the highway.

After identifying UCAs and creating corresponding safety constraints, it is necessary to verify Step 1 results to ensure that there are no conflicts between safety constraints or requirements. To help engineers finish this task, the authors develop a prototype user interface based on the meta-model in Figure 6. Table 2 gives an example of a conflict existing between two safety constraints derived from UCAs. As can be seen, "Rule" 3 conflicts with "Rule 8", that is, Rule 3 requires ADS to provide motion planning when the automated vehicle is moving in the left lane on the highway and the driver is present, while Rule 8 asks ADS not to provide motion planning when the vehicle is moving on the highway under extreme weather conditions (e.g., storm).

Integration of STPA Step 2 for Creating Functional Safety Requirements

After creating safety constraints for validating safety goals for the motion planning and acceleration commands, engineers can perform STPA Step-2 to:

- Create functional safety requirements for each safety goal.
- Assign responsibilities to system components for defining the preliminary architecture of the automated vehicle.

The meta-model provides the traceability necessary for performing these tasks. Consider UCA 9 -Not providing motion planning is undesired if the automated driving system is in active mode, while the vehicle is en route and within operational boundary. Causal scenarios and factors related to this UCA include:

Scenario 9: The automated vehicle is moving on the highway and a lane-change is needed because of newly detected events or obstacles. But it does not change lanes because the automated driving system does not provide motion planning in time. The automated controller believes that the vehicle is not in the correct route or within the operational boundary.

- Causal factor 9.1: mission details are incomplete.
- Causal factor 9.2: unintended mode changes from autonomous drive mode to manual mode.
- Causal factor 9.3: automated driving system believes that the automated vehicle is not en route because map data was modified by unauthorized access.
- Causal factor 9.4: automated driving system believes that the automated vehicle is not in the operational boundary because map data (e.g., route direction) or timing information (e.g., clock) was modified by unauthorized access.

While the first two causal factors are recognized as safety concerns, the third and fourth are related to cyber vulnerabilities in the given architecture described before. This can be understood by an example involving the "reversible lane" that is designed to allow the traffic to move in either direction based on certain conditions [19]. It is designed to avoid traffic jams during rush hours, as shown in Figure 8. As can be seen, the yellow lane in the middle is designed to be "reversible," and its traffic direction depends on the time of a day. Traffic moves from east to west (left) in the morning while moving in the opposite direction (right) in the evening. Currently, the lane direction is indicated by control signals LED signs or physical separation. In the future, when automated vehicles support V2I communication, such design can be achieved if the automated driving system receives route information from road units through vehicular network, such as DSRC.

One problem is that V2I communication is subject to cyber-attacks. For the vehicle architecture that only allows autonomous vehicles to move within the "operational design domain" (e.g., time, geographical condition, etc.), an adversary who can modify the map data related to the highway direction or spoof the timing information (e.g., vehicle clock) can easily compromise the normal operation of the vehicle. Consider that the vehicle is moving in the reversible lane from east to west, but the timing information that the automated driving system receives suggests that it is in the evening. This is the case where the automated driving system finds itself "out of the operational design boundary" because it can only move the vehicle from east to west in the reversible lane during the morning, according to the traffic rules. To make things worse, if the automated driving

FIGURE 8 An example of UCA-9 caused by cyber vulnerabilities.

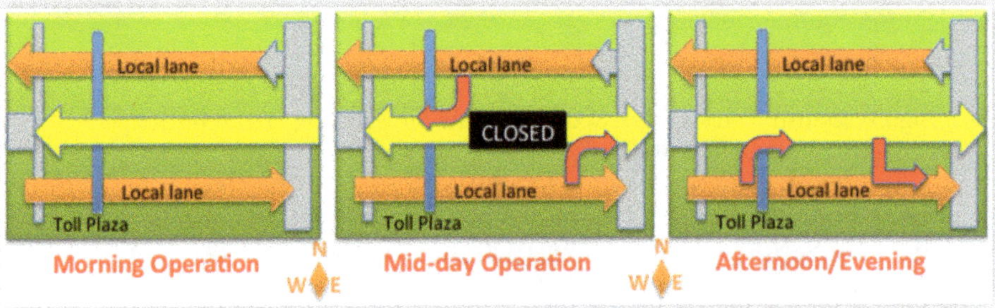

FIGURE 9 Traceability for creating FSRs and making architectural decisions.

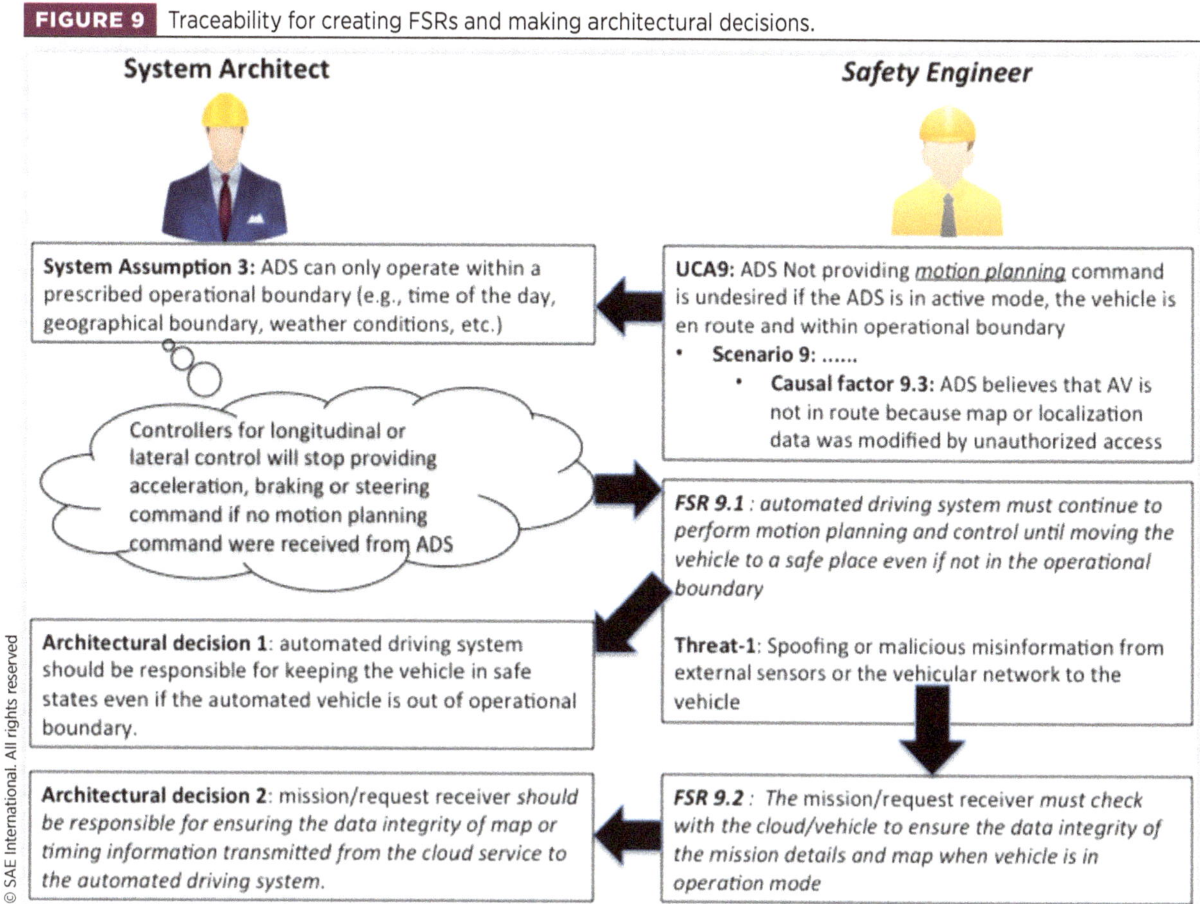

system is not designed for moving the vehicle to a "safe" place when the vehicle is not en route or is out of operational boundary, it will not stop providing motion planning and control commands, putting the automated vehicle in a dangerous situation (no one is controlling the vehicle).

Based on the scenarios and causal factors identified above, functional safety requirements can be created to define the vehicle's architecture. Figure 9 illustrates how FSR 9.1 is created by considering the worst-case scenario. According to the system assumption that an automated vehicle can only operate within the operational boundary, ADS will stop providing motion-planning command under the scenario because of causal factors 9.3. Engineers may then decide to assign the responsibility of keeping the vehicle moving until it is in a "safe" zone to ADS, as illustrated by Architectural decision 1.

In addition to deriving functional safety requirements, scenarios and causal factors identified in STPA Step-2 also provide guidance in identifying cyber-security threats. For example, a cyber-security threat (Figure 9) related to scenario 9 and causal factors 9.3 and 9.4 can be

- Threat-1: Spoofing or malicious misinformation from external sensors or the vehicular network to the vehicle

Although Threat-1 is framed as a security concern, it also has safety implications, as spoofing attacks on the map data or vehicular network can also result in vehicles violating minimum separation distance (H-1) or not following the intended pathway (H-2) that is originally planned by the automated driving system. Engineers can create FSR 9.2 to specify the required behavior of the mission/request receiver-the mission/request receiver must check with the cloud/vehicle to ensure the data integrity of the mission details and the map when the vehicle is in operation mode. Figure 9 only illustrates how STPA Step-2 results can be used to derive functional safety requirements and responsibility assignment, rather than specifying a user interface for performing this task.

As guidelines for engineers when making architectural decisions, consider several guidance questions after creating functional safety requirements for the automated driving system.

- Given the system assumption that the automated driving system should only operate within its operational boundary (e.g., time, geographical and route conditions, weather conditions, etc.), which module should keep the vehicle safe when the automated vehicle is still moving but the automated driving system is deactivated because it is out of the operational boundary?
- If the automated driving system is responsible to move the vehicle to a "safe" place even if it is out of the operational boundary, what test cases are necessary to ensure the system is devoid of hazardous behaviors?
- If another controller is required to operate the vehicle when the automated driving system is not within the operational boundary due to unexpected events, what safety requirements should be created such that control commands from the new controller do not conflict with the old one's?

Consideration for Integration with Cyber Security Analysis

Safety assurance in ISO 26262 does not consider cyber security issues that are the focus of the cyber-security guidebook for automotive systems-SAE J3061 [21]. One of the goals of providing the cyber-security guidebook is to "evaluate threat analysis and risk assessment (TARA) methods using a simple approach to allow effective implementation across the automotive industry." [22] STPA can complement the threat analysis in the sense that it can find missing threats or evaluate existing ones, rather than substituting TARA.

Figure 10 illustrates how Threat-1 derived from STPA Step-2 results. The case study on an automated driving system starts with high-level hazards and losses that must be prevented (i.e., two vehicles violating minimum separation distance) and two functions for the automated driving system-motion planning and acceleration, rather than explicitly considering potential threats or malicious attacks to the automated vehicle. After finishing Step 1 and Step 2 with all causal scenarios and factors identified for UCAs, engineers are able to create not only functional safety requirements, but also security requirements for system components (e.g., the automated driving system and the mission/request receiver). Also, for the message that is used by the automated driving system, such as route direction in the map data or timing information, engineers can decide its criticality based on which function (e.g., motion planning) is using it and the ASIL rating of the safety goal for that function.

FIGURE 10 Relation diagram for functional safety, STPA and cyber-security.

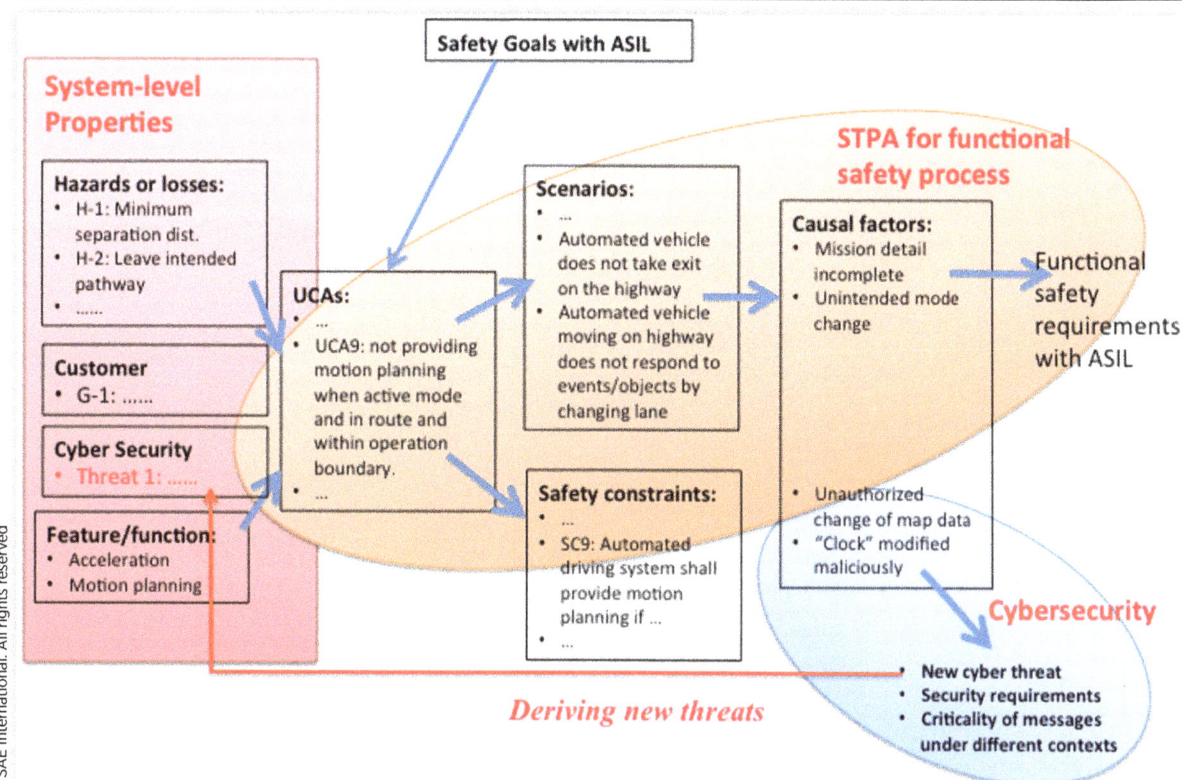

Summary/Conclusion

This paper describes a process map for integrating STPA into the functional safety process based on ISO 26262. Specifically, three steps in the process map are illustrated through a case study on an automotive system.

- System assumptions and components from item definition are used to form the system engineering foundations for STPA.
- UCAs identified and safety constraints created in STPA Step 1 are used to evaluate existing safety goals with ASIL ratings developed from HARA.
- Causal scenarios and factors for UCAs identified in STPA Step 2 help engineers create functional safety requirements and make architectural decisions.

For the modeling and tool support, a meta-model based on SysML is developed. In addition to enabling the integration of STPA with ISO 26262 process, the paper also shows how the meta-model can be used to deal with system-level properties other than safety early in the concept phase, i.e., identifying undesired control action related to customer experience in STPA Step-1 and derive new cyber security threats based on scenarios and causal factors from STPA Step-2. It is worth mentioning that STPA is an iterative process and can also be used to create technical safety requirements and generate test cases. But those aspects are not covered because the focus of the paper is on functional safety requirements.

Definition/Abbreviations

STPA - System Theoretic Process Analysis
SC - Safety Constraint
CF - Causal Factor
ADS - Automated Driving System

References

1. Flemming, C., *Safety-Driven Early Concept Analysis and Development* (Cambridge, MA, 2015).
2. International Standardization Organization, "ISO 26262- 1:2011(en) Road vehicles - Functional safety - Part 1: Vocabulary," International Standardization Organization.
3. Leveson, N., *Engineering a Safer World* (Cambridge, MA: MIT Press, 2012).
4. Hommes, Q., "Safety Analysis Approaches for Automotive Electronic Control Systems," https://www.nhtsa.gov/sites/nhtsa.dot.gov/files/2015sae-hommes-safetyanalysisapproaches.pdf.
5. Mallya, A., "Using STPA in an ISO 26262 Compliant Process," *Proceedings on Computer Safety, Reliability, and Security: 35th International Conference, SAFECOMP 2016*, Trondheim, Norway, Springer, September 21-23, 2016.
6. Thomas, J., Sgueglia, J., Suo, D., Leveson, N. et al., "An Integrated Approach to Requirements Development and Hazard Analysis," SAE Technical Paper 2015-01-0274, 2015, doi:10.4271/2015-01-0274.
7. Placke, S., Thomas, J., and Suo, D., "Integration of Multiple Active Safety Systems using STPA," SAE Technical Paper 2015-01-0277, 2015, doi:10.4271/2015-01-0277.
8. Abdulkhaleq, A. and Wagner, S., "XSTAMPP: an eXtensible STAMP Platform as Tool Support for Safety Engineering," 2015.
9. Becker, C. and Hommes, Q., Transportation Systems Safety Hazard Analysis Tool (SafetyHAT) User Guide (Version 1.0). No. DOT-VNTSC-14-01, 2014.
10. Leveson, N., "A New Accident Model for Engineering Safer Systems," *Safety Science* 42, no. 4 (2004): 237-270,
11. Object Management Group, "The OMG System Modeling Language Version 1.4 specification," 2015, http://www.omg.org/spec/SysML/1.4/.
12. Thomas, J., "Extending and Automating a Systems-Theoretic Hazard Analysis for Requirements Generation and Analysis," Ph.D dissertation, Cambridge, MA, 2013.
13. Suo, D., "Tool-Assisted Hazard Analysis and Requirement Generation Based on STPA," Master thesis, Cambridge, MA, 2016, http://hdl.handle.net/1721.1/105628
14. Suo, D. and Thomas, J., "An STPA Tool," *3rd STAMP/STPA Conference*, Cambridge, MA, 2014.
15. Leveson, N., "Completeness in Formal Specification Language Design for Process-Control Systems," *Proceedings of the Third Workshop on Formal Methods in Software Practice*, ACM, 2000, 75-87
16. Van Eikema Hommes, Q., "Review and Assessment of the ISO 26262 Draft Road Vehicle - Functional Safety," SAE Technical Paper 2012-01-0025, 2012, doi:10.4271/2012-01-0025.

17. Young, W., and Leveson, N. G., "An Integrated Approach to Safety and Security Based on Systems Theory," *Communications of the ACM*, 57, no. 2 (2014): 31-35.
18. Hommes, Q., "Assessment of Safety Standards for Automotive Electronic Control Systems," Report No. DOT HS 812 285, National Highway Traffic Safety Administration, Washington, DC, 2016, June.
19. https://en.wikipedia.org/wiki/Reversible_lane.
20. Bell, R., "IEC 61508: Functional Safety of Electrical/Electronic/Programme Electronic Safety-Related Systems: Overview," *Computing & Control Engineering* 11, no. 1 (1999): 5/1-5/5.
21. SAE International Surface Vehicle Recommended Practice, "Cybersecurity Guidebook for Cyber-Physical Vehicle Systems," SAE Standard J3061, Iss, Jan. 2016.
22. SAE International, "SAE Committee Busy Developing Standards to Confront the Cybersecurity Threat," Automotive Engineering Magazine article, http://articles.sae.org/13809/.
23. Ujiie, R., "Using STPA in the Design of a New Manned Spacecraft," *The 2nd STAMP workshop*, Cambridge, MA, 2013.

CHAPTER 10

The Development of Safety Cases for an Autonomous Vehicle: A Comparative Study on Different Methods

Junfeng Yang, Michael Ward, and Jahangir Akhtar
Birmingham City Univ.

The Connected and Autonomous Vehicles (CAVs) promise huge economic, social and environmental benefits. The autonomous vehicles supposed to be safer than human drivers. However, the advanced systems and complex levels of automation could also bring accidents by tiny faults of hardware or errors of software. To achieve complete safety, a safety case providing guidance on the identification and classification of hazardous events, and the minimization of these risks needs to be developed throughout the entire development lifecycle process of CAVs. A comprehensible and valid safety case has to employ appropriate safety approaches complying with the automotive functional safety requirements in ISO 26262. The technical focus of present work is on the comparative study of different safety approaches, in particular, Failure Mode and Effects Analysis (FMEA) method and Goal Structuring Notation (GSN) method that have been employed to generate lists of hazardous events, safety goals and functional safety requirements at the vehicle level. A case study on the safety case development of INISIGHT autonomous vehicle has been carried out using the aforementioned methods. This case study covers the safety argument of battery and charging system that supply the whole electric power for INSIGHT vehicle. The safety of this systems has been assessed along with their potential for malfunction together with the layers of protection. The results and conclusions from case study analyses suggest the safety case of CAVs can be developed in a highly effective manner by employing a combined method of GSN and FMEA.

CITATION: Yang, J., Ward, M., and Akhtar, J., "The Development of Safety Cases for an Autonomous Vehicle: A Comparative Study on Different Methods," SAE Technical Paper 2017-01-2010, 2017, doi:10.4271/2017-01-2010.

Introduction

Rapid growth in personal transport is frightening in terms of the spiraling number of injuries and deaths, global pollution and climate change. Back in 2009, 5.5 million accidents in the USA, involving 9.5 million vehicles, killed ~34k people and injured >2.2M others, including 240k hospital admissions [1]. In addition, cars and trucks are estimated to cause 20% all U.S. CO_2 emissions [2]. For the exploding numbers of cars in the developing world, the statistics are even more terrifying. The CAVs equipped with more sensors to detect other road uses and pedestrians, and much higher levels of computer control promise huge reductions in accidents, congestion, and pollution. For example Google claim their driverless car could reduce accidents by 90%, wasted time and fuel by 90%, and massively increase the utilization of cars, meaning fewer cars overall [3]. With a huge market worth in view, every major car manufacturer in the world is developing CAVs. One estimate for sales of autonomous vehicles is 95 million per year by 2035 [4]. The IEEE predicts that autonomous vehicles could be as much as 75% of the market by 2040 [5]. The automotive industry makes a substantial contribution (>£60Bn) to the UK economy, and is expected to see considerable growth in the next decade and more.

One of the earliest reports on autonomous vehicle appeared on 1948 which concerned the development of cruise control in vehicles [6, 7]. Since then the work has been developed by many researchers to include areas such as mechanical antilock braking, electrical stability control, laser based cruise control, pre-crash mitigation. Some of the first autonomous car projects in the 1980s were the Navlab (1980) and the ALV (Autonomous Land Vehicle) in 1984 that were organized by Carnegie Mellon University (CMU). They have continued to develop the autonomous car since then. Recently the CAVs industry keeps blooming and many companies including Mercedes-Benz, General Motors, Continental Automotive Systems, Autoliv Inc., Bosch, Nissan, Toyota, Audi, Volvo, Google and Tesla have developed autonomous cars [8, 9].

Figures for UK's CAVs development show a similar trends. Within UK, the Centre for Connected & Autonomous Vehicles, CCAV, has been established to help ensure UK's world leadership in developing and testing connected and autonomous vehicles. Since 2015, CCAV has continuously funded a series of projects, e.g. GATEway, Venturer, UK Autodrive, INSIGHT, i-MOTORS and FLOURISH, on intelligent mobility research and development. Among them, the INSIGHT project aiming to develop driverless shuttles with a particular focus on improving urban accessibility for disabled and visually-impaired people will be thoroughly introduced. And the safety case developed for the INSIGHT pod will be discussed as the case study in the follow in g sections.

The INSIGHT [10] project is a collaborative project to develop existing autonomous vehicles for safe, slow speed (max 15 MPH) operation in pedestrian areas and pavements, with connectivity not only to control and manage the vehicles, but also for innovative data collection and presentation applications that interact with users and other customers of the systems. An existing electric connected & autonomous vehicle design [11] has been upgraded with advanced sensors to detect and recognize pedestrians, cyclists, mobility scooters, and other vehicles on adjacent roads. These detection capabilities will enable more advanced decision making and a more nuanced approach to way finding, and a smoother ride rather than the simple start/stop common in such systems today. The general view of INSIGHT autonomous vehicle can be seen in Figure 1.

The INSIGHT pod vehicle is driverless and self-steering (autonomous driving SAE Level 5 [12]), electrically-powered light-weight vehicle designed to carry up to four people and their luggage (including pushchairs and bulky items), see Figure 2 for the general interior view. While the INSIGHT pod is suitable for almost any age group, it has been designed with inclusive at its heart. The vehicle has wheelchair access and, INSIGHT will look specifically at its use by the elderly and those who need assistance in transport, for example the visually impairment. The pod will not just assess the physical passenger experience, such as internal comfort and safety, but also the supply journey information such as calling response times, destination, connections and other support information, all delivered by a human voice interface.

FIGURE 1 General exterior views of the INSIGHT vehicle.

FIGURE 2 General interior views of the INSIGHT vehicle.

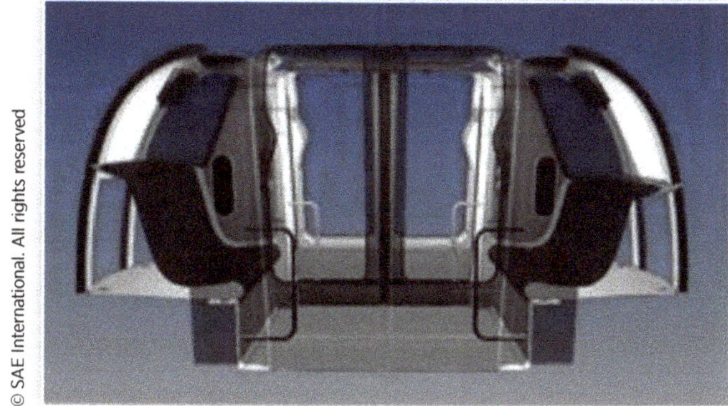

The project activities require a safety case to be made before commencing in order to provide assurance that any reasonable residual risks have been minimized and where possible avoided all together. In addition, a safety case based upon the road vehicle functional safety standards is also required to demonstrate that the vehicle can be safely and reliably driven. The typical approaches documenting safety cases include textual format, tabular form e.g. using FEMA, graphical notations, e.g. GSN. All these approaches have been employed to formulate the safety cases for various autonomous vehicles. For instance, LUTZ pathfinder automated vehicle [13] has developed a defendable safety case using FEMA approach together with a tailored application of ISO26262 automotive functional safety standard [14], and concluded the use of human intervention is required for trials. Another similar project, ULTra CAVs providing personal rapid transition between the T5 Business Car Park and Terminal 5 of Heathrow Airport generated a safety case using a combined method of FEMA and GNS. The ULTra CAVs has been safely running since 2011 and delivered in excess of 3.5million passengers, which provide a strong and convincing evidence on that successful safety case.

The INSIGHT vehicle is based upon an existing design ULTra CAVs, where the vehicle supposed to operate in an unconstrained pedestrian area instead of is confined to a well-defined purpose built track. It is necessary therefore that an appropriate safety is developed for INSIGHT vehicle. The present work seeks to explore various approaches for developing a safety case for INSIGHT vehicles. These approaches will incorporate with the general safety management follows a diverse set of legislation and guidance, e.g. SAE J3018 and J3061 [15], UK Code of Practice for Testing Driverless Car [16] and the Road Traffic Act [17]. The performance of various approach will be compared and analyzed.

Vehicle Layout and ISO26262

Vehicle Control System and Propulsion System

In order to allow level 5 autonomous vehicle operation, an autonomous control system has been developed for INSIGHT pod vehicle. This autonomous control system consists

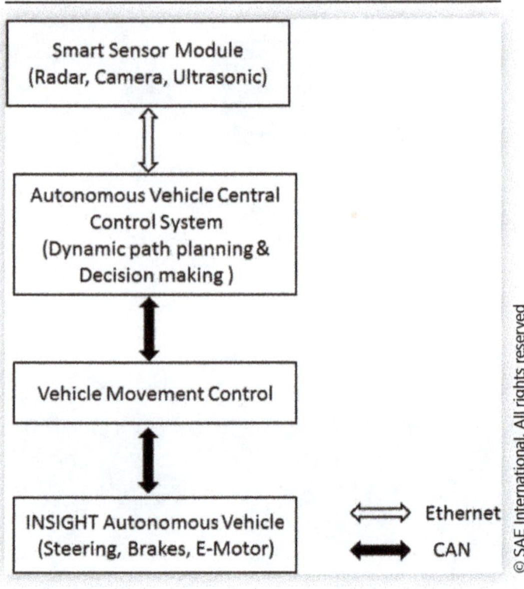

FIGURE 3 Basic functional topology of an INSIGHT vehicle control system.

of situational awareness system, central control system facilitating dynamic path planning and decision making, and vehicle movement control system. The smart sensor module ((a group of sensors, e.g. long- and short-range radars, front-, rear-, and side- stereo-cameras and ultrasonic sensors)) connected via wired Ethernet to autonomous vehicle central control system to improve path planning and decision making. The module integrates steering, brakes and E-Motor which respond to demand from the vehicle control system, transmitted via a CAN bus. Figure 3 presents the basic control system architectural of an INSIGHT vehicle. Note that this basic control system is expressed exclusive of the environmental monitoring sensor system, human-machine-interaction control and 4D tactile system for a clearer view.

The INSIGHT vehicle has two Li-ion battery units, high voltage (48V) electric power for traction power system, and low voltage (24V) electric power for vehicle control system and door actuation system. The typical propulsion system for an INSIGHT vehicle is shown in Figure 4. As can be seen, the motor control module converts the 48V DC battery power into low voltage 3 phase AC power while simultaneously controlling motor torque speed and direction. The AC E-Motor drives the vehicle through the front wheels via a fixed ratio transmission and a differential mounted in a transaxle. Note that, the nominal system voltage is restricted to 48 volts to minimize the shock risk.

ISO26262 Road Vehicle Functional Safety Standard

ISO26262 is an adaption of IEC 61508 [18] to meet specific needs of automotive industry. It is the first comprehensive standard that addresses safety related automotive systems comprised of electrical, electronic, and software elements that provide safety related functions. It seeks to address the following important challenges in today's road vehicle technologies: the safety of new electrical and electronics hardware and software

FIGURE 4 Typical propulsion system schematic for an INSIGHT vehicle.

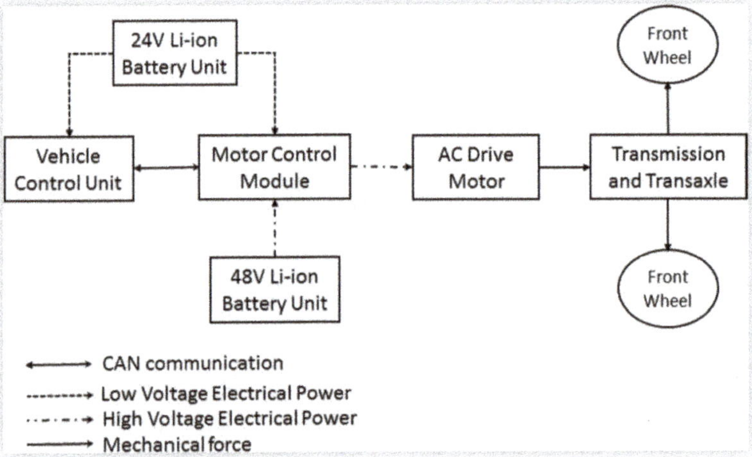

FIGURE 5 System lifecycle approach to safety: used throughout a safety case.

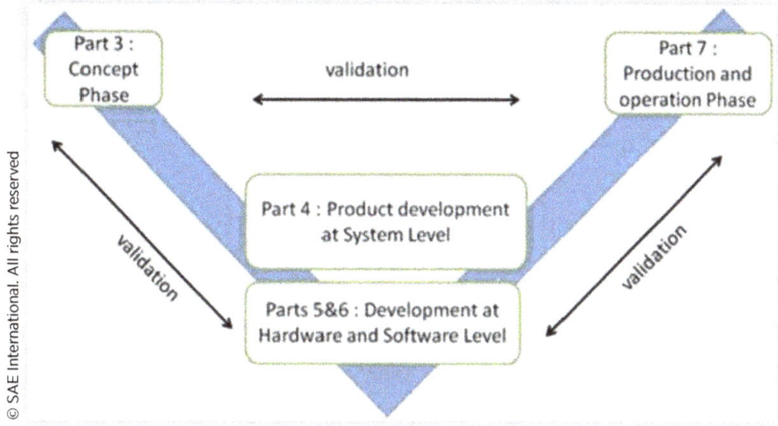

functionality in vehicles; the trend of increasing complexity, software content, and mechatronics implementation; the risk from both systematic failure and random hardware failure. It also provides guidance on how to avoid risk in creating safety-critical systems and regulates critical testing processes.

ISO 26262 defines a safety case as an "argument that the safety requirements for an item are complete and satisfied by evidence compiled from work products of the safety activities. Figure 5 present a system lifecycle approach (V-model) used throughout a safety case. This lifecycle model represents the development of the system from first concept to operation. The concept phase (Part 3) refers to the initial big picture of autonomous vehicle in terms of styling and functionality, etc. Parts 4-6 refers to the vehicle develop and software/ hardware development. Part 7 refers to the final product. Validations refer to various trials, e.g. commissioning test (vehicle shakedown), on-road tests of hardware/software, and trial under a public pedestrian area at different phases. The V-shape is due to the fact that the testing and verification steps are performed in reverse order from design and implementation.

Failure Model and Effects Analysis Method

Failure Mode and Effects Analysis (FEMA) method developed initially for analyzing malfunctions of military systems [19] uses a structured, systematic spreadsheet to documents all the possible failures, risk assessment and management strategies in a design, a manufacturing or assembly process, or a product or service.

The typical FMEA spreadsheet captures all systems/components information including Items, Functional Requirements, Failure Modes, and Causes of Failure. Each possible Cause of Failure has an associated Risk to it which is derived from its Occurrence & Severity. After this the first focus is on Design actions, which means going through the Causes of Failure and mitigate these risks down to the lowest possible level. After the design actions have been considered, the part should be ready for validation through either physical testing or rationale, proving its robustness and ability to meet the safety performance requirements. During Validation, design review, customer reviews/testing issues/concerns may be raised. These issues or failures need to be fed into FMEA ensuring that additional risk is added to the parts of concern proving that we have mitigated that failure and are fit to continue the validation process. The example spreadsheet of FMEA method was given below

TABLE 1 Example risk assessment represented using FMEA spreadsheet

Item	Function Requirement	Failure Model	Casual Factor	Immediate Consequences	Severity	Frequency	Exposure	Hazard Ranking	Mitigation
Battery	supply electric power for E-motor	supply insuffcient electric power	over-heating; degradtion	vehicle lose power	3	2	2	6	BMS monitory battery voltage and temperature

Taking the advantages of intuitively clear and high viability, FMEA method has been further developed and adopted by the aerospace and automotive industries.

Goal Structuring Notation Method

Goal Structuring Notation (GSN) method [20] is a graphical notations for the representation of arguments, which was first proposed by T. P. Kelly [21] under the inspiration of Toulmin's argument model [22]. GSN method employs a simple notation of argumentation structures that have been proven to be effective for provides objective safety evidence, therefore has been widely used for developing safety cases for the industrial use and research purpose. Recently, GSN method has been incorporated into ISO26262 to satisfy the critical safety assurance of automotive systems, e.g. start/stop system, and EPS system [23, 24, 25, 26, 27, 28].

Typically, GSN method consists of a group of symbol of notations linked by directional arrows explicitly representing the individual elements (safety goals, solution, context and strategies) of an argument and the relationships between these elements, such as rectangular boxes for safety target, ovals for assumption or justification, circle for evidence (solution), parallelogram-shaped boxes for strategy (argument), rounded-end boxes for context (additional information). An example safety argument constructed using GSN is given below.

As shown in Figure 6, the safety goal of targeted system needs to be claimed by identifying the possible hazards and mitigating them through sufficient and appropriate evidences. Due to the complexity of system, the top-level safety goal usually has to be decomposed into sub-level goals and this decomposition may continue until sub-level claim and evidences asserted.

FIGURE 6 an example safety argument represented using GSN.

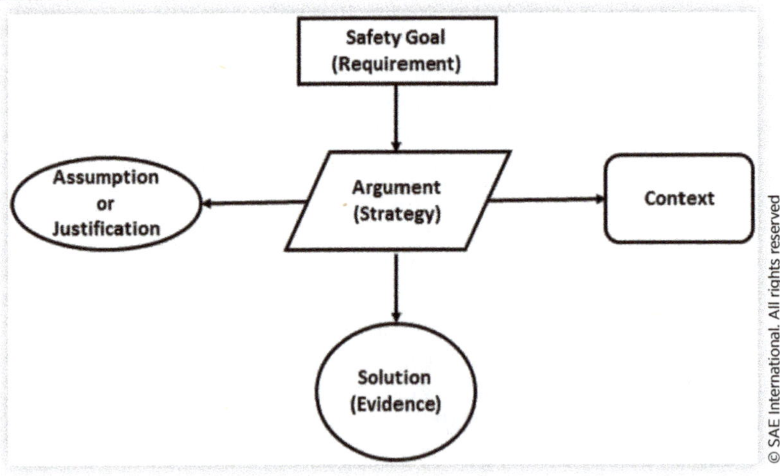

Both FMEA and GSN have been widely used for the risk assessment of automotive industries, and proven to be valid and capable methods. The following sections provide a detailed description on a safety case for INISHGT autonomous vehicle constructed using FMEA and GSN methods incorporating with ISO26262 road vehicle functional safety standard. In addition, the performance appraisal of these method provides guidance toward a valid and defendable safety case.

Safety Case Development

In the present work, the safety case investigates and documents all the hazards and risks associated with autonomous vehicle, including mechanical system, electric hardware, application and embedded software, communications, health and safety of passengers, roads users, pedestrians, risk assessments, safety systems of work and insurance and liability.

A valid safety case for an autonomous vehicle consists of four main inter-dependent components, namely:

- Safety target that must be addressed to assure vehicle safety.
- Evidence for the safety target obtained from study, analysis and test of the vehicle system.
- Argument showing how the rationale indicates compliance with the safety target.
- Context identifying the basis for the argument presented.

A set of safety targets for the vehicle commissioning tests is generated with the objective of achieving acceptable safety considering the prototype nature of INSIGHT vehicle. Based on the goals of the safety work, the principles of safety process rationale argument were chosen as:

1. Hazard Generation. Identifying the vehicle operational situation and the possible hazardous events associated with safety targets.
2. Risk Assessment. Classifying each hazard in terms of frequency of occurrence, severity of resulting harm and controllability of hazard, and determining the Automotive Safety Integrity Level (ASIL) of system by considering the SAE J2980 standard [29].
3. Hazard Management. Addressing safety requirements through an appropriate combination of system design in accordance with the ASIL indicated.

An ASIL shall be determined for each hazardous event based on its severity level (S1-S3), probability of exposure (E1-E4) and controllability level (C1-C3) in accordance with Table 2. The number 1 represents lowest level and 4 the highest one. The classification of severity, exposure and controllability are given in SAE J2980.

As can be seen, Four ASILs are defined: ASIL A, ASIL B, ASIL C and ASIL D, where A representing the least stringent level and D the most stringent level. QM indicates quality management system can be sufficient to develop element(s) that implement the safety requirement allocated to the intended functionality. Or it can support the rationale for the independence between the intended functionality and the safety mechanism.

Figure 7 summarizes the safety process rational argument employed in the present work. If a system has high ASIL and subjects to the constraints,

TABLE 2 ASIL Determination per ISO26262:2011

ASIL Classification		C1	C2	C3
S1	E1	QM	QM	QM
	E2	QM	QM	QM
	E3	QM	QM	A
	E4	QM	A	B
S2	E1	QM	QM	QM
	E2	QM	QM	A
	E3	QM	A	B
	E4	A	B	C
S3	E1	QM	QM	A
	E2	QM	A	B
	E3	A	B	C
	E4	B	C	D

FIGURE 7 Safety process rationale argument.

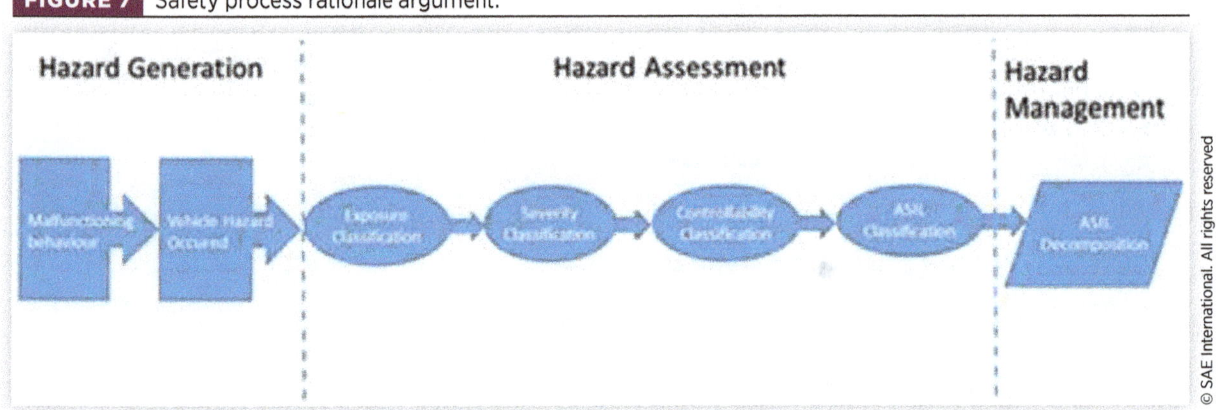

its safety requirement can be decomposed by multiple redundant subsystem working together, each with a lower ASIL. This process is so-called ASIL decomposition that allows the best safety strategies to be developed efficiently.

Case Study

The INSIGHT autonomous vehicle has two Li-ion battery units mounted on to the battery tray of the rear. The battery and charging system supply as one of the most important systems supply the whole electric power for the INISIGHT vehicle. These batteries incorporates an on-board Battery Management System (BMS) in the vehicle. The BMS has the feature of measuring cell voltage and temperature, performing cell balancing function and monitoring the cell fault conditions, providing these information to the external systems via CAN. Since the Lithium-ion batteries contain flammable electrolyte and may pose a fire/explosion and other hazards when it's overheated or short-circuited. Thereafter the safety case must include the risk assessment on battery and charging system. This section takes the battery and charging system as the case study to examine the product-based safety rational argument.

Table 3 present the FMEA-based safety argument for the battery and charging system. The potential malfunctioning behavior relevant to the battery and charging system and potential hazards have been identified. Then severity level, controllability and probability of exposure of this vehicle hazard and ASIL level are classified for each hazards. Failure of the battery and charging system is clearly undesirable. However system design features mitigate the risk associated with such failures to acceptable levels. Figure 8 shows GSN-based product argument structure for the battery/charging Systems Safety argument. This valid argument (strategy) is supported by evidence in compliance with ISO 26262.

As can be seen clearly, FMEA is better for documenting the evidence and context. But it seems hardly to convey all necessary information effectively along. While, GSN are better for presenting a decision making process, especially the decomposition of safety goal on a complex system which involves a number of risks. However, for a complex system, the GSN has to cover a number of sub-goals which may cause an intricate GSN structure and make it difficult to follow.

TABLE 3 FMEA-based safety argument for the battery and charging system of INSIGHT vehicle

Entry id	Item	Cause of failure	Consequence	Mitigation	Controllability	Severity	Frequency	ASIL Ranking	General Comments
1	Battery/ charger	Electric current leakage	Electric shock	Nominal system voltage is restricted to a non-lethal level (48volts). No passenger exposure to high voltage. No circuits used within the passenger compartment operate at voltages above 24V. Vehicle charging contacts are mounted under the vehicle and are inaccessible to passengers. Neither vehicle nor the mounted charging contacts are live when they are not connected together	C1	S3	EI	QM	
2	Battery/ charger	Resistive connection	Overheating	Temperature sensing employed to detect excessive temperature at charging contacts. Purpose designed connectors used for all high current connections.	C1	S1	E3	QM	
3	Battery/ charger	Rapid charging or discharging	Fire/explosion	Battery protection fuse mounted within battery pack. The battery pack is physically separated from the passenger compartment by bulkheads.	C1	S3	E3	A	
4	Battery/ charger	Battery heating	Fire/explosion	Battery chargers automatically control charging to avoid overcharging. Vehicle controller continuously monitors individual battery voltage and temperature; it will stop charging if overcharging or over temperature fault conditions are detected.	C1	S3	E2	QM	

FIGURE 8 GSN-based Safety goal rational argument for a battery/charging system of INSIGHT vehicle.

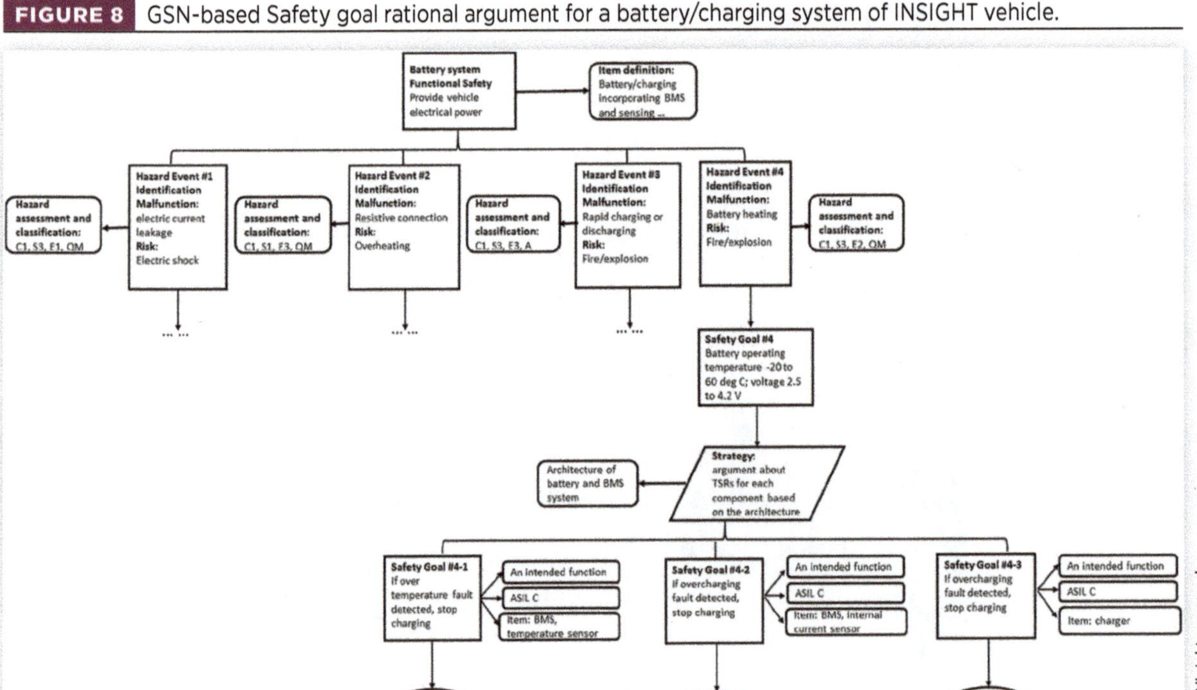

Conclusions

Safety has been a prime consideration throughout the development of the INSIGHT autonomous vehicle. The present work has developed a safety case in order to provide assurance that any reasonable residual risks have been avoided during the vehicle commissioning test. This safety case builds in accordance with the ISO26262:2011 road vehicle functional safety standard. The principle conclusion of safety assessment is that no features of the INSIGHT design concept and operating concept would indicate that the level of safety of the INSIGHT system would be unacceptable. The INSIGHT vehicle is therefore acceptably safe to commence operations.

A diverse approach to the assessment of the safety of the system has been adopted, including assessment against FMEA, GSN and SAE guidance, and quantified risk assessment. The risks associated with all the identified hazards are considered to be in the tolerable or acceptable risk categories. A case study on a battery and charging system of INSIGHT vehicle demonstrates the typical structure of safety goal rational argument. The analysis results indicated that the approach adopted was appropriate.

Regarding the performance appraisal of FMEA and GSN methods, it found FMEA is better for documenting the evidence and context. But it seems hardly to convey all necessary information effectively along. While, GSN are better for presenting a decision making process, especially the decomposition of safety goal on a complex system which involves a number of risks. Hence, a combined method of FMEA and

GSN is suggested to contracture a valid and defendable safety case in an efficient and effective manner.

In terms of future work, we would like to continue our safety case to explore the aforementioned method to address the unique aspects of CAVs, e.g. navigation system and decision-making system, which have not been thoroughly discussed.

Contact Information

Junfeng Yang, PhD
School of Engineering and the Built Environment
Faculty of Computing, Engineering and the Built Environment
Birmingham City University
City Centre Campus
Millennium Point
Birmingham B4 7XG
United Kingdom
Phone: +44 (0)121 300 4293
Junfeng.Yang@bcu.ac.uk

Acknowledgments

This project was funded by Innovate UK (grant agreement No. 102583) and supported by the Centre for Connected and Autonomous Vehicles, UK.

Definitions/Abbreviations

CAV - Connected and Autonomous Vehicle
FMEA - Failure Mode and Effects Analysis
GSN - Goal Structuring Notation
ASIL - Automotive Safety Integrity Level
BMS - Battery Management system

References

1. http://www-nrd.nhtsa.dot.gov/Pubs/811363.pdf.
2. http://www.ucsusa.org/our-work/clean-vehicles/car-emissions-and-global-warming.
3. http://www.google.co.uk/about/careers/lifeatgoogle/self---driving-car-test-steve-mahan.html.
4. http://www.navigantresearch.com/research/autonomous---vehicles.
5. http://www.ieee.org/about/news/2012/5september_2_2012.html.
6. "keesler news," keesler, 1 1 1948, [Online] accessed October 13, 2016, http://www.keesler.af.mil/AboutUs/FactSheets/Display/tabid/1009/Article/360538/history-of-keesler-air-force-base.aspx.
7. Ioannou, P.A. and Chien, C.C., "Autonomous Intelligent Cruise Control," *IEEE Transaction on Vehicle Technology* 42, no. 4 (1993): 657-672.

8. Thorpe, C., Hebert, M., Kanade, T., and Shafer, S., "Toward Autonomous Driving: The CMU Navlab," *IEEE* (1991): 31-41.
9. Thorpe, C., Hebert, M., Kanade, T., and Shafer, S., "Vision and Navigation for Carnegie-Mellon Navlab," *IEEE* 10, no. 3 (1988): 362-373.
10. Centre for Connected and Autonomous Vehicles, UK, INSIGHT Project, http://insightcav.com, January 2016.
11. Heathrow Pod, accessed July 2015, http://www.ultraglobalprt.com/wheres-it-used/heathrow-t5/.
12. SAE International Surface Vehicle Information Report, "Guidelines for Safe On-Road Testing of SAE Level 3, 4 and 5 Prototype Automated Driving Systems (ADS)," SAE Standard J3018, Iss, Mar. 2015.
13. Peters, A., "Safety of the LUTZ Pathfinder Automated Vehicle," *22nd ITS World Congress, Paper number ITS-2427*, Bordeaux, France, October 5-9, 2015
14. ISO 26262 - Road Vehicles -Functional Safety. Parts 1 to 12.
15. SAE International Surface Vehicle Information Report, "Guidelines for Safe On-Road Testing of SAE Level 3, 4 and 5 Prototype Automated Driving Systems (ADS)," SAE Standard J3018, Iss, Mar. 2015.
16. UK Department for Transport, "The Pathway for Driverless Car: A Code of Practice for Testing, 2015.
17. UK Government, "Road Traffic Act," 1991.
18. IEC 61508 - Functional Safety of Electrical/Electronic/Programmable Electronic Safety-related Systems.
19. United States Department of Defense, "MIL-P-1629 - Procedures for Performing a Failure Mode Effect and Critical Analysis," Department of Defense (US), MIL-P-1629, November 9, 1949.
20. Goal Structuring Notation Working Group: GSN Community Standard Version 1. http://www.goalstructuringnotation.info/, 2011.
21. Toulmin Stephen, E., *The Uses of Argument* (Cambridge University Press, 1958).
22. Kelly, T.P., "Arguing Safety - A Systematic Approach to Safety Case Management," DPhil Thesis, YCST99-05, Department of Computer Science, University of York, UK, 1998.
23. Kelly, T. and Weaver, R., "The Goal Structuring Notation-A Safety Argument Notation," *Proceedings of the Dependable Systems and Networks 2004 Workshop on Assurance Cases*, July 2004.
24. SAE International Surface Vehicle Recommended Practice, "Cybersecurity Guidebook for Cyber-Physical Vehicle Systems," SAE Standard J3061, Iss, Jan. 2016.
25. Palin, R. and Habli, I., "Assurance of Automotive Safety: A Safety Case Approach," *SAFECOMP 2010*, Vienna, Austria, 2010.
26. Palin, B., Ward, D., Habli, I., and Rivett, R., "ISO 26262 Safety Cases: Compliance and Assurance," *IET Intl. System Safety Conf.*, 2011.
27. Habli, I., et al., "Safety Cases and Their Role in ISO 26262 Functional Safety Assessment," *32nd International Conference on Computer Safety, Reliability, and Security*, Toulouse, France, 2013.
28. Matsuno, Y., "D-Case Ediotor," http://www.il.is.s.u-tokyo.ac.jp/deos/dcase/.
29. SAE International Surface Vehicle Recommended Practice, "Considerations for ISO 26262 ASIL Hazard Classification," SAE Standard J2980, May 2015.

epilogue

Juan R. Pimentel
Professor of Computer Engineering
Kettering University

In this book, we have discussed the ISO standard 26262 in the context of the safety of automated vehicles. Although the second edition of ISO 26262 (2018) included many enhancements to the original edition of 2011, it did not include automated vehicles. As noted in the epilogue of book 3 in this SAE collection, there are two primary approaches to reduce risk, a management approach and a technical approach. The former is based on using systems engineering (SE) methodologies, particularly the V-model, while the latter involves specific risk reduction measures or mechanisms such as redundancy. In this book we have summarized the main concepts of SE as applied to automated vehicles, emphasizing the concept of MBSE (model-based systems engineering) and the use of the SysML language. It is not claimed that SysML is the most complex or best language; however, it is one that is often used in automotive and automated vehicle applications. We have included a set of ten papers that are representative of research in this category. Although there is a fair amount of work in various underlying areas such as SE, control systems, DFMEA, STPA, HARA, and risk reduction techniques, more work is needed to unify these areas into a coherent body of knowledge with explicit applications to automated vehicles.

It is noteworthy that STPA is a relatively new methodology (at least when compared to DFMEA) that explicitly uses SE principles and control systems structures to identify hazards and to design mitigation solutions for risk reduction. As demonstrated in some of the papers included in this collection, STPA can benefit immensely from MBSE and SE languages such as SysML. Nevertheless, the STPA process is somewhat ad hoc, and more work is needed to define a possible international standard. Such standard would greatly contribute to the application of STPA for automated vehicles.